"A meticulously documented analysis of India's decarbonization possibilities by one of our most thoughtful and experienced administrative scholars. Essential reading for those interested in India's future and its impact on the world."

Suman Bery
Vice Chair, NITI-Aayog (National Institution for Transforming India)

"India has made a vital commitment to achieve net zero greenhouse gas emissions by 2070, and put in place some of the specific policies such as renewable power investment required to get there. But to ensure rapid progress at least possible cost India needs a detailed road map of the actions required in all sectors of the economy. Ajay Shankar's excellent book provides that road map and argues persuasively for the policies required to make it achievable. He shows that in attaining the net zero objective—certainly by 2070 and quite likely even earlier—India would also become a globally competitive green industrial economy. *India's Energy Transition* is analytically robust but also visionary, and a crucial guide to how India can both become a rich developed nation and make a major contribution to avoiding the potential catastrophe of global warming."

Lord Adair Turner
Chair, Energy Transitions Commission

RISING AMBITION

Carving new pathways

INDIA'S ENERGY TRANSITION

AJAY SHANKAR

THE ENERGY AND RESOURCES INSTITUTE
Creating Innovative Solutions for a Sustainable Future

ISBN 978-93-94657-19-9

First reprint, 2026

Published by

The Energy and Resources Institute (TERI)
Darbari Seth Block, India Habitat Centre
Lodhi Road, New Delhi 110 003, India
Tel.: +91 11 2468 2100 or 2468 2111
Fax: +91 11 2468 2144 or 2468 2145
Email: mailbox@teri.res.in | Web: www.teriin.org

To my wife and lifelong friend, Meera, whose support and encouragement has made this book see the light of day.

ACKNOWLEDGEMENTS

I would like to begin by appreciating TERI and especially its Chairman, Nitin Desai and its Director General, Dr Vibha Dhawan. TERI has given me the ambience and the intellectual space to think and write.

A big thank you to my coauthors of the TERI papers on the key dimensions of India's Energy Transition: Taruna Idnani, Meenu Saini, Robin Mazumdar, and Sanchit Aggarwal. These papers were written after the COP26 announcement that India would achieve net zero by 2070. They led to the crystallization of my ideas which are set out in the book.

Girish Sethi, AK Saxena, and their enthusiastic teams, who work on energy and industry, have been of great help.

A special thank you to Arpita E. Victor. Her painstaking and diligent assistance in finding information and references, and in the revisions of the manuscript has been invaluable. Lakshmi Subramanium has been such a cheerful and willing personal aide.

I am grateful to the members of the Publishing Solutions team of TERI, notably Sachin Bhardwaj, Rajiv Sharma, and Vijay Nipane for their support that made publishing of this book at such a remarkable speed possible.

I would like to thank my nephew, Pranav Sarin for designing the book cover, as a personal favour.

I must acknowledge my debt to the extraordinary ethos of confidence and the 'can do' spirit of the Indian

Administrative Service (IAS) in which I had the privilege of serving and experiencing how good policy and administration is 'the art of the possible'.

The Ministers, Secretaries, and colleagues with whom I served in government, all helped me in better appreciating the complexities and challenges of development in India.

I must express my gratitude to my friends, peers, experts in energy and industry leaders with whom I have interacted and from whom I have learnt so much over the years.

CONTENTS

Introduction

At COP26 in November 2021 in Glasgow, India announced a substantial increase in its ambition on climate change. Speaking at the Summit,[#] Prime Minister Modi made the following key declarations:

> "First – India will take its non-fossil energy capacity to 500 GW by 2030.
> Second – India will meet 50% of its energy requirements from renewable energy by 2030.
> Third – India will reduce the total projected carbon emissions by one billion tonnes from now till 2030.
> Fourth – By 2030, India will reduce the carbon intensity of its economy by more than 45%.
> And fifth – By the year 2070, India will achieve the target of Net Zero."

The goal of achieving 500 GW of fossil fuel-free capacity by the end of the decade, 2030, took most observers, in India as well as in the rest of the world, by surprise. To get a sense of perspective, India had at that time a total generating capacity of only 383 GW, out of which the non-fossil fuel capacity was barely 159 GW. This huge increase in fossil fuel-free capacity would have to come primarily from renewables.

India for the first time announced its intention of becoming net zero by 2070. China had by then stated that they would become net zero by 2060. Given the huge gap between where China had reached in its development journey and where India was, an extra 10 years to get to net zero was only reasonable.

---

https://www.mea.gov.in/Speeches-Statements.htm?dtl/34466/National+Statement+by+Prime+Minister+Shri+ Narendra+Modi+ at+COP26+Summit+in+Glasgow

We, in TERI, were like everyone else, taken by surprise by the announcements in Glasgow. As a think tank, TERI, since its inception 50 years back, has been engaged in analysing key issues and challenges relating to energy and the environment. We decided to think through the pathways that India would need to create to get to the ambitious goals that it had set for itself, the immediate goal of creating 500 GW fossil fuel-free capacity by 2030 and then of becoming net zero by 2070.

We came up with a series of discussion papers, held roundtables, and engaged with stakeholders and experts. These have helped in the crystallization of views and refinement of ideas.

This book on India's energy transition is the result. It gives a holistic and comprehensive picture of how the journey has been unfolding so far. It describes the creative pathways that resulted in early successes and generated confidence for increase in ambitions. The first two chapters give an account of successes and rising confidence. The subsequent chapters describe the more difficult challenges that lie ahead in making the transition to net zero. These chapters are more conceptual and prescriptive. This is unchartered territory with so many uncertainties. It will be a long journey over a few decades.

India has faced with huge challenges. At the turn of this century, its foremost energy challenge was to complete the huge task of taking electricity to all its villages and homes. India then had the largest population in the world living without electricity, more than in Sub-Saharan Africa. Then over 80% of households were using biomass, fuelwood, and cow dung cakes for cooking. Power cuts were the norm due to endemic electricity shortages. This was holding India back. A qualitative transformation in the power supply situation was imperative. India was able to achieve the major and essential energy transition of taking electricity to all homes by around 2020. It was also able to create adequate power-generation capacity at the same time. The programme of providing clean cooking gas connections and stoves to all households has been completed. New policies and creative programme designs were crafted for attempting these historic transitions. Progress was faster and smoother than expected. In the process confidence naturally rose. This experience is set out in the first chapter.

India had been an early mover in looking at alternative sources of energy. After early success in developing wind power, an ambitious National Solar Mission was launched in 2010. The Mission began to do far better than initially planned. The success in renewables has gone far beyond initial expectations. Ambition and innovative smart programme design made this possible. Renewable energy capacities have been created by private investment with hardly any outgo of funds from the budget. This has been an unusual achievement for India. This is described in the second chapter.

This success in the Solar Mission began to generate increasing confidence and rising ambitions. India was able to move from its traditional position and become a willing partner to the Paris Accord of 2015. The large Nationally Determined Contributions (NDCs) intimated by India for 2030 were achieved ahead of time. The major announcements at COP26 in November 2021 displaying greater climate leadership followed.

Renewable energy generation is inflexible and variable. It creates new challenges. As solar power is the main source of renewable energy for India and electricity is generated only when the sun shines, rapid development of large-scale grid storage of electricity is a necessity. The imperative of storage forms the subject matter of the third chapter. Large-scale grid storage options of pump storage projects (PSPs), concentrated solar projects (CSPs), and battery energy storage systems (BESSs) are looked at. Some fresh ideas for creating storage rapidly are put forward. These break new ground. India is now set to become one of the early movers in creating storage projects, without which a carbon-free electricity system is not feasible.

In the fourth chapter the more challenging journey from the creation of 500 GW of fossil fuel-free capacity to creating a carbon emission-free electricity system is discussed. The contours of this critical transition are set out. The major policy choices in the coming years that would need to be considered along with their implications are analysed. The key milestones in this transition would be, first, the decision not to build new thermal plants. Emissions from electricity generation would then peak. The phased retirement of all the thermal power plants would then has to take place for a carbon emission-free electricity system to emerge. These would have to happen when electricity demand would be rising as the Indian economy would continue to grow. Closure of thermal plants would entail substantial costs. Ways of coping with these are discussed.

In the fifth chapter 'Entering the Hydrogen Economy' the tougher challenge of decarbonizing the hard-to-abate sectors using green hydrogen is analysed. Hard-to-abate sectors are those that cannot be electrified. Sectors which can be electrified would become carbon free as electricity generation becomes carbon free. Road transport is a good example of an energy-intensive sector which is being electrified. Success in the hard-to-abate sectors is essential for reaching net zero. For these, green hydrogen is being seen as a potential solution. India has become an early mover in trying to move towards the new green hydrogen economy with its well-funded Hydrogen Mission. The challenges, possibilities, and pathways to getting to the global frontier are suggested. As resources are a constraint, affordability has been a central consideration in the ideas put forward. Setting up of pilot plants using green hydrogen in each downstream sector is suggested. Technical capacities would be created along with the discovery of the actual higher costs. A rational analysis on inter-se priorities and sustainable, least-cost pathways would then become feasible.

In the concluding chapter, a holistic view of the journey to net zero is presented. The major choices that would lie ahead in the coming years and decades along with their implications are analysed. It is argued that a serious discussion on the policy choices in the future should begin now to avoid taking sub-optimal decisions which would cost more in the long term. It would be a mistake to take the view that the goal of achieving net zero is so far away that consideration of feasible pathways is, at present, premature. The goal of net zero and the costs of alternative pathways need to be factored in by firms as well as government institutions while considering lumpy, long-term capital-intensive investments. Getting policy decisions right is never easy. Becoming net-zero appears daunting at the present. But this national goal has to be achieved.

The belief that India can and will succeed is the underlying premise of the book. Further, this transition would facilitate the journey to becoming a developed nation. There would be substantial costs, but these are unavoidable for achieving net zero. They may become lower with the right policy choices. The benefits in becoming a globally competitive green economy may well turn out to be higher.

This is an India-centric book. It is conceptual and analytical. It is written for the general reader as well as for the experts and its focus is on policy choices. Though these are complex and difficult, the attempt has been to put them simply and be readily comprehensible. The hope is that it would enrich the policy discussion on this fundamental transition that is unfolding. It avoids technical details, computations, and modelling studies. There are no headline figures for the costs of the transition. The attempt has been to look at least-cost pathways to net zero. Discussion and a broad consensus need to emerge around these. The sooner the better. Needless to add, these would evolve with experience, technological progress, and changes in costs.

India is so large, home to over 1.4 billion people. Its sheer size makes its energy transition of interest to the larger global climate community. Being a developing country, its experience would also be of relevance for other developing nations facing similar situations and challenges.

To the extent India succeeds it would make a substantial difference towards the achievement of net zero by the world.

Note: The views in the book are solely those of the author.

CHAPTER 1
Achieving Universal Access

SDG (Sustainable Development Goal) 7: "Ensure access to affordable, reliable, sustainable and modern energy to all."

Introduction

India has succeeded in achieving this important SDG.

At the beginning of this century, the power sector was amongst India's major challenges. Shortages in generating capacity were endemic and power cuts were common. Those with resources took care of themselves by buying personal diesel generators. The work of extending electrification in rural areas was practically at a standstill. More than half of the rural population did not have access to electricity. Regional disparities were widening. Bringing clean modern cooking energy to rural India was not even under consideration.

In two decades, a radical transformation took place. India had created adequate power-generation capacity. Almost all homes had got electricity. The programme of providing clean cooking gas to all was well under way. It has now been completed.

How did this happen? Power had been a high priority in India's Five-year Plans. [1,2,3] It was always getting a large share of the financial resources for planned development. What made such extraordinary success possible in these two decades? There were fundamental changes in approaches, new policies were put in place, and innovative programme designs were crafted. A new law, the Electricity Act, 2003,[4] replaced the earlier laws regulating electricity. The central government assumed direct responsibility for the completion of rural electrification with a new national programme. This is a fascinating story that is described in this chapter.

Completing Electrification

According to the 2001 national census, 44% of the population in India did not have access to electricity.[5] At around 400 million this was a larger number than those without access to electricity in Sub-Saharan Africa. Out of a total of around 6,00,000 villages in the country around 1,25,000 were unelectrified.[6] Many states had completed village electrification by the 1970s.[7] The unelectrified villages were all concentrated in the states in the lower Ganges Valley and adjoining areas, with 80,000 of them being in the two most populous and least-developed contiguous states of Uttar Pradesh and Bihar.[8] In Bihar, about 40,000 out of 45,098 villages remained unelectrified.[9] A few thousand villages were being reported as having been de-electrified—either the wires had been stolen or the villages were never electrified. The prognosis was bleak. Parliament was informed in an answer to a question then that Bihar would, at the existing rate of village electrification, take over 700 years to electrify all its villages.[10] The International Energy Agency (IEA) in its report around that time had stated that India would be able to provide electricity to all its people sometime between 2040 and 2050, with 2040 being the optimistic scenario. The states with the large number of unelectrified villages were being left behind in economic development. Not surprising, as without electricity economic activity in these villages remained in the pre-industrial age. There began a growing recognition that this had to change. Completing rural electrification needed to be given much higher priority as a development goal.

This marked a major shift, with the central government assuming responsibility along with the state governments to provide electricity access to all. Electricity was originally the responsibility of the state governments, being a state subject under the Constitution. The Electricity (Supply) Act was enacted in 1948, the first major legislation enacted by independent India. It was modelled on a similar legislation nationalizing the electricity supply industry in the United Kingdom (UK) by their first Labour Party government which was voted into power after World War II. State Electricity Boards (SEBs) were established in India along the lines of the Central Electricity Generating Board (CEGB) of the UK. Prior to independence, India had private licensees who supplied electricity in major cities. Private entrepreneurs took licences for cities where the supply of electricity was profitable. The rest of India remained in darkness. At the time of independence in 1947, only 1500 villages had access to electricity. The newly created State Electricity Boards were given the responsibility to extend the supply of electricity to all unserved areas. They gradually took over the private licensees and only four survived; two in Mumbai, one in Kolkata, and one in Ahmedabad. The State Electricity Boards were given a large share of the Plan funds of the state governments.

A structural problem began in the early days. In the rates the State Electricity Boards charged their customers, they had a differentiated structure, charging the well off more than the cost of supply. This enabled them to charge less than the cost of supply from the weaker sections and farmers. This regime of 'cross-subsidy' was in effect a tax on the consumption of the 'rich'. The 'rich' kept declining in relative numbers as the supply of electricity kept expanding. This cross-subsidy model ran into difficulties as the subsidizing consumers shrank relative to the rapidly growing category of subsidized consumers, especially in the states which saw relatively lower rates of industrial and economic growth. The financial health of the State Electricity Boards began to deteriorate. The issue of the poor financial health of the sector persists (Annexe 1.A.). There are ongoing efforts to improve the financial health of the Distribution Companies (DISCOMs). However, electricity being the essential service that it is, state governments and the central government have no option but to provide bailouts to keep the lights on. They have been doing so repeatedly.

The states of north, west, and southern India which were doing well had completed the electrification of all their villages by the 1970s. The states of central and eastern India were not doing well economically, and the pace of their rural electrification programme was relatively slow. They just did not have the financial resources for faster progress.

For critical development challenges the central government had from the early years after independence been assuming responsibility. It had been implementing programmes, providing funds, and technical support to the states in the fields where, under the Constitution, the states were responsible. The achievement of self-sufficiency in food grain production, agriculture being a state subject, and maintaining it for a rapidly increasing population since the 1970s is the most important example of success. Major programmes in the areas such as health and nutrition are also financed and driven by the central government.

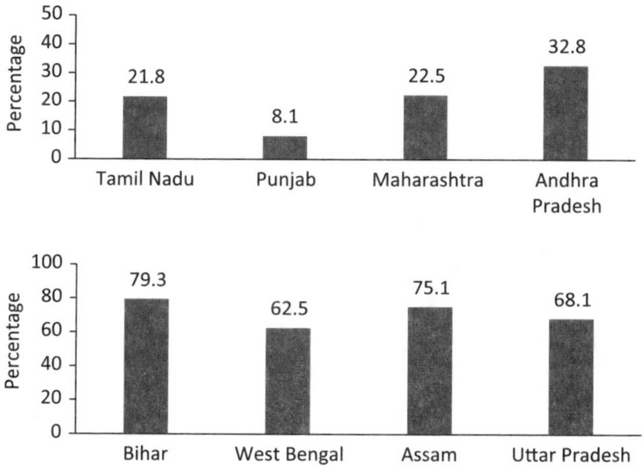

Figure 1 Percentage of households without access to electricity in 2001 [11]

In the power sector, to help the states overcome their acute power shortages in the early 1970s, the central government set up thermal and hydropower-generating companies as central public sector undertakings. The Constitution was amended to bring electricity from 'the state list' to the 'concurrent list' to give the central government responsibility for electricity along with the state governments. The National Thermal Power Corporation, NTPC, was set up with assistance from the World Bank.[12] It has achieved exemplary benchmarks for timely execution of projects, smooth and efficient operations, plant availability, and profitability. It has grown to become one of the largest power-generating companies in the world.[13] As a group it has a total generating capacity of almost 75,000 megawatts (MW) with a stand-alone thermal capacity of nearly 59,000 MW. It accounts for about 25% of the total electricity generated in India.[14]

The view that the central government had to act for faster rural electrification began to get traction. After all, in the 21st century such a large part of rural India could not be allowed to stay without electricity. Life without electricity was literally a life in the pre-industrial age. The central government assuming responsibility is explicitly reflected in Section 6 of the new Electricity Act, 2003, which states:

"The concerned state government and the central government shall jointly endeavour to provide access to electricity to all areas including villages and hamlets through rural electricity infrastructure and electrification of households."

The initial effort, however, turned out to be too feeble. In 2002 a programme in the form of a 4% interest subsidy on the loans that the State Electricity Boards would take for rural electrification was put in place.[15] This failed to have an impact as the State Electricity Boards in the states where the programme was most needed were in extreme financial distress. Taking large loans on the scale needed for the gigantic task of completing rural electrification did not turn out to be viable. The programme did not take off.

An alternative strategy was reflected in the Electricity Act, 2003. It created the enabling regulatory framework for the full and free play of market forces in rural areas to meet the huge unmet demand through the given provision:

"where a person intends to generate and distribute electricity in a rural area to be notified by the state government, such person shall not require any licence for such generation and distribution of electricity (Section 14)."

The generation and supply of electricity in rural areas was made a free unregulated business. Anyone could set up a supply business and decide whom to give electricity to, and at what price. The market mechanism also failed. It did not even begin to deliver. Giving market forces full freedom was expected to lead to a surge in the private supply of electricity in rural areas for those who could afford to pay. Their numbers were large enough to constitute a huge market. This lack of success is a sobering experience for those who have great faith in the power of free markets. The supply side in the market does not always appear instantaneously in the real world and move to meet the demand that exists. This is experienced repeatedly in developing countries in many areas. The state has no option but to act in such cases of 'market failure' if it would like to change the status quo quickly.

Gradually, democratic politics began to drive rural electrification higher up as a major development goal. In the 2004 national election the BJP (Bhartiya Janata Party) which was leading the ruling coalition promised in its election manifesto [16] that it would electrify all villages by 2009 and all households by 2012. The Congress Party, which was in the opposition, in its manifesto[17] which was released later promised to take electricity to all households in five years. The election results came as a surprise and the Congress Party formed a coalition government of the UPA (United Progressive Alliance).

Even if the Congress Party may have made the promise to take electricity to all in haste and without thinking through the implications, ignoring their election promise was not an option for them. Within the government there were influential voices which had serious reservations about the central government taking on such a big responsibility and finding the money for it. What was even more important, where would the electricity to supply these villages come from when the country was already facing power shortages. But the government of Prime Minister Manmohan Singh did decide to launch a national programme for completing rural electrification.[18] The big transformative decision was to provide 90% of the cost of electrifying villages as a grant by the central government. The balance 10% was to be given as a loan to the state distribution utilities (DISCOMs). The loan, though only 10%, facilitated the sense of ownership of the programme by the state DISCOMs. Under the Electricity Act the DISCOMs had the universal service obligation of providing electricity on demand to any consumer in their licence area. This commitment of 90% funding by the central government made the completion of rural electrification possible.[19, 20]

With funds no longer being a constraint, the ability to implement such a large programme became the key challenge. The states where the maximum work had to be done were the ones with relatively weak state capacity for efficient and timely execution of programmes. The programme design had major innovative features in departure from tradition which made speedy and better implementation possible. These were:

a. Generally, financial assistance for any programme from the central government goes to the state governments. They implement the programme using this money. In a radical departure, funds for this programme did not go to the state governments. The programme was implemented through the government-owned development financial institution (DFI), the Rural Electrification Corporation (REC).*

b. The implementation of the programme was undertaken through specific projects. Projects for extending the grid for the electrification of all villages in a compact area were to be prepared and then sanctioned by the REC. These projects had to be implemented through turnkey fixed price and time contracts. This too marked a departure from the traditional mode of implementing projects by state agencies. By giving full responsibility of execution to the contractor, he had a clear incentive to finish the project within time. The projects were large and therefore better firms responded to the bids for execution of the projects. New benchmarks for timely execution of projects were created. Funds were sanctioned for individual projects and disbursements were linked to implementation milestones and onward payment to the execution agency.

c. Further, the public sector undertakings of the Ministry of Power were also deployed to execute projects in some states, notably Bihar, where it was felt that the state government agencies did not have the capacity for the

* The REC had been set up in 1969 to provide loans to the State Electricity Boards to energise pump sets in rural areas for irrigation and higher food production. India had been dependent on food imports since independence and becoming self-sufficient in food production was a very high priority. India did achieve its 'Green Revolution' in a few years, became self-sufficient in foodgrain production by the early 1970s and has remained so since. It even exports some rice. The credit it provided to the State Electricity Boards was used to extend the supply of electricity in rural areas primarily for irrigation but also for village settlements. However, when the State Electricity Boards in the weaker states lost their capacity to take loans and service debt for rural electrification, the programme slowed down. The responsibility for implementing the new national programme for completing rural electrification was given to the REC.

execution of such a large programme on their own. These central organizations had a culture of efficient project management. Though these projects were outside their normal domain, they delivered timely outcomes.

d. The definition of village electrification was tightened. At least 10% of the households of a village had to have electricity connections for it to be declared as having been electrified. Reporting of names of villages being electrified on a central portal was introduced. This ensured that reporting was correct even in states with weak governance.

States readily agreed to these major departures from normal practice as they could see that this implementation design had the potential for speedy execution, with all the money coming from the central government and transforming their rural areas.

Trying to electrify over 1,25,000 villages out of a total of 6,00,000 in the country was a truly gigantic task (Annexe 1.B.). The programme took off and succeeded beyond expectations. By 2012, about 1,00,000 villages were electrified.[21] Clearly, India had acquired the project execution capacities for implementing such an ambitious programme.

In its *Energy Access Report* in 2017,[22] the IEA highlighted India's achievement:

"Half a billion people have gained access to electricity in India. This remarkable growth puts India on course to achieving access to electricity for all in the early 2020s—a colossal achievement."

After villages were electrified, it was found that the pace of households taking electricity connections was slow.

After almost all villages had been electrified, India still had about 240 million people without electricity access in 2016. To complete household electrification, the central government decided to complete household electrification in a mission mode and provided financial assistance under the Saubhagya programme. Under this, provision was made for, "free metered connection for economically poor households and a charge of Rs 500 (adjusted in electricity bills in 10 monthly instalments) for households other than poor" (Annexe 1.C.).[23] By the beginning of the 2020s, the mammoth task of electrifying all households was completed.

Around 700 million people had got access to electricity in about 15 years. [24] This was unparalleled in the history of electrification. India was transformed.

Overcoming Power Shortages

In the early 2000s, those who had access to electricity experienced frequent power cuts and blackouts, more in some states and less in others. [25] The creation of additional generating capacity did not keep pace with demand. Shortages were perennial. This was the case despite the central government's NTPC growing and increasing capacity rapidly. India had, as a result, a rapidly growing market for diesel generating sets for supply of backup electricity.[26] From industrial and commercial establishments to government offices to residential apartments and private homes, they were all buying diesel generating sets. The situation was highly unsatisfactory. But it had been like this for years.

In 2000, the government led by Prime Minister Atal Bihari Vajpayee decided that India had to move beyond doing more of the same. It needed to think creatively and act to achieve a fundamental transformation of the power sector. It decided that a new law should replace the existing laws regulating the electricity sector. It took the view that it would be worthwhile to make the effort to evolve a broad-based consensus, a wise step. This was also the era of coalition governments at the national level. This approach facilitated the evolution of a legislation which suited Indian conditions and resulted in highly positive outcomes.

After three years of intense, protracted, and detailed consultations with all stakeholders, a new law, the Electricity Act, 2003, replaced the three existing laws which governed the sector. These were the Indian Electricity Act, 1910, the Electricity (Supply) Act, 1948, and the Electricity Regulatory Commission Act, 1998. In an unusual bipartisan consensus, this was supported by all political parties in the Parliament except the small Marxist parties. Its salient features are summarized in Annexe 1.D.

The new law consciously chose not to follow the policy prescription coming from many quarters of emulating the UK model of complete price deregulation, the separation of the supply business from the wires business, and the creation of full retail competition. The American firm Enron, the major foreign company in the power sector in India at that time, was a major proponent of this. This 'reform' had been undertaken in California. The near collapse of the electricity system in California with spiralling prices led the Governor there to make a huge financial intervention in the 'market'. Prices stabilized. Enron, which had taken forward positions assuming that very high prices would be maintained, faced huge losses. Later Enron collapsed. It ceased to exist. There were useful lessons of the dangers of adopting an extreme doctrinaire view of the benefits of a free deregulated market in electricity.

The new Electricity Act was evolved to suit Indian conditions. Its intention as reflected in the preamble was: "to consolidate the laws relating to generation, transmission, distribution, trading and use of electricity—promoting competition therein, protecting interest of consumers and supply of electricity to all areas—promotion of environmentally benign policies."

Full competition was created in generation where, as in all production, the potential gains from competition are the maximum. This was combined with having DISCOMs with universal service obligation and regulated tariffs. Large consumers were given the freedom to opt out and buy power directly. And all were free to set up their own captive power plants, individually or collectively ("a person may construct, maintain or operate a captive generating plant and dedicated transmission lines" [Section 9]).

The distribution licensee had to ensure reliable supply for the rising demand of all the consumers in his area. This was done by the DISCOM entering into long-term power purchase agreements (PPAs). These agreements became the basis for new investment in generation. With the security of committed long-term sale of the output at a price which gave stable and adequate returns on the capital invested, raising money both for equity and debt became feasible. Competitive procurement of power through such long-term contracts became the preferred mode of getting new investment into generation, gradually moving away from the earlier practice of government-owned entities putting up power plants and getting cost-based tariffs for the electricity they generated. The Electricity Act provided for acceptance by the Regulatory Commissions of the tariff from competitive procurement. Section 63 mandates that the Commission "...shall adopt the tariff if such tariff has been determined through a transparent process of bidding."

The earlier momentum of integrating state grids was accelerated and an all-India synchronous power grid emerged. This made possible the creation of an all-India market and investment in very large power projects, with attendant economies of scale at one site, from which the electricity generated would be sold to DISCOMs in many states. This system of de-risking of new investment in generation reduced the risk premia and enabled easy financing of increasing private investments in generation. This introduced competition which in turn incentivized cost reduction. The consumer benefited from stable electricity prices. The potential of large spikes in prices characteristic of other fully deregulated markets was not there in the market design in India. The UK, the pioneer in 'reforms' of the electricity sector with a fully deregulated spot market saw consumer electricity prices rise by 54% under the impact of the Ukraine conflict.[27, 28] The full benefits of competition in generation (production), which is where the maximum gains are likely, were obtained and passed on to the consumers. The DISCOMs through this system of long-term contracts were able to ensure reliable supply and stable prices for their consumers.

The positive transformation that this new system brought about went far beyond initial optimistic expectations. After the enactment of the new law in 2003, it took three to four years for the new regime to settle down and begin working. The central government prepared guidelines and bid documents for competitive procurement of power through a long-term contract from new power stations. This was undertaken for the first time in India. Two very large projects of competitive procurement of 4,000 MW each by a group of state DISCOMs were steered by the central Power Ministry.[29] Investor response was good, and the market-determined tariff turned out to be lower than expectations. These set the pattern for competitive power procurement by the DISCOMs. There was a surge in private investment in generation, going far beyond what most analysts expected. Private investors were able to raise money from domestic and global financial markets. By 2016 at a macro all-India level new investment in generation had made the power supply position comfortable. The share of the private sector in new capacity creation rose sharply and was larger than that of the public sector. This was far more than the most optimistic expectations.

Earlier efforts after the economic liberalization of 1991 to get private investment into power generation had not succeeded.[30] In 1992, India made a concerted effort to get foreign investment into power generation. The rate of return on equity of 12% for power generation, which had been set for NTPC on the advice of the World Bank, was raised to 16% in the currency of the foreign investment. The central government even agreed to provide counter guarantees to supplement the state government guarantees for the first set of eight fast-track projects. The major learning of this unsuccessful effort in the decade of the 1990s was that it may have been better to first get the Indian private sector to invest in generation competitively as happened later. Overeagerness may have led the foreign investors to seek higher margins assuming the country risk premium to be greater. There was also the natural incentive to gold plate capital costs in a non-competitive situation where returns were based on capital costs. Enron was the first investor setting the benchmark. It did set up its plant. The tariff turned out to be unsustainably high. After Enron collapsed, this plant was taken over by Indian government undertakings. It is still a white elephant. Only three of the fast-track projects were put up. None of the other proposed plants, though there was a gold rush type of situation with 189 offers for more than 75 GW capacity addition by mid-1995, came up as the tariffs were turning out to be too high and unsustainable. The World Bank concluded that this initiative to get private investment into generation 'turned out to be a dud overall'. India now has a competitive domestic private sector in power generation. Competition on tariffs through long-term power purchase agreements has delivered. There is assured capacity and stable low prices over the long duration of the power purchase agreements. The transformation is illustrated in Figure 2.

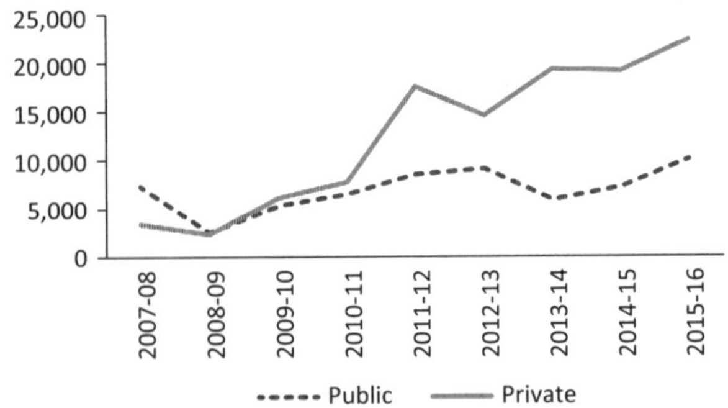

Figure 2 Sector-wise annual installed power capacity addition (2007–16) (Annexe 1.E.)

Nevertheless, providing reliable quality supply of electricity round the clock to all consumers is still a work in progress. Even when generation capacity at the national level is adequate, the upgradation of the local distribution network has, in many areas, not kept pace with increasing demand, resulting in avoidable power cuts (Annexe 1.F.). Going forward, DISCOMs would have to ensure that new-generation capacity is created to meet increasing demand. This will remain a challenge in the coming decades.

Demand growth is a consequence of GDP growth and the constituent elements of this growth in terms of energy intensity. Uncertainties are inherent in projections of GDP growth rates and the energy intensity of the sectors which would drive growth. Services-led growth in India has resulted in relatively modest growth in the demand for electricity. But, if manufacturing takes off, electricity demand growth rates would become much higher. Planning for resource adequacy to provide reliable supply has to accept the uncertainty inherent in demand projections. It, therefore, needs to be dynamic and flexible. The creation of reserve capacities to meet demand when it grows more than expected is essential. Decision-makers in the power sector must accept this principle and take decisions accordingly.

Clean Cooking Energy

The provision of clean cooking energy makes a remarkable difference to the quality of the lives of women. The time spent in collecting firewood or preparing cow dung cakes, the biomass energy sources for cooking of the preindustrial era, is large, a few hours every day, at least. Besides, burning of biomass for cooking indoors has an adverse impact on the lungs and the eyes of women who do the cooking. It is also a major source of air pollution.

Provision of clean cooking energy deserved higher priority. It is not that efforts were not made. For decades attempts were made to design and make smokeless cookstoves using biomass. Success was claimed and demonstrated. But switchover from the traditional modes of using biomass on a significant scale did not happen for a variety of reasons. Over 80% of households continued to use traditional biomass till 2015.[31] Almost all rural households were using biomass for cooking. In the cities clean cooking gas, liquefied petroleum gas (LPG) cylinders, were being provided by the public sector oil companies. The provision of piped gas supply in some cities had also started. This had been a more recent development.

The decisive transformation began with Prime Minister Modi starting the Pradhan Mantri Ujjwala Yojana (PMUY) in 2016. The central government funded this programme under which a free gas cooking stove and cooking cylinder along with the gas connection were given to women of below the poverty line (BPL) rural households by the public sector oil marketing companies (OMCs) of the central government (Annexe 1.G.).[32] The supply network for LPG cylinders has been extended into rural areas at remarkable speed. Under PMUY about 103.3 million households had got connections by the end of 2024. There has been an increase in LPG coverage from just 62% in 2016 to near saturation.[33] The Indian Parliament was recently informed of this achievement (Annexe 1.G.).

The complete switchover to clean cooking and giving up the use of biomass altogether must be achieved. The gas cylinder is expensive and its affordability for poorer households is therefore an issue. For them, finding money for a refill is a challenge though a subsidy of Rs 200 is being provided for the refill of a cylinder. In recent state elections in India, promises of providing cylinders at subsidized and affordable rates were made. Democratic politics is likely to keep the issue alive. The refill of a cylinder may have to be made cheaper for households of the weaker sections.[34]

The government may choose to give a higher subsidy for the refills. If providing money from the Budget is difficult, the government could impose a cess on the sale of petrol, diesel, and gas and raise the resources for the subsidy. Or, it could ask the petroleum companies to go back to the old ways of cross-subsidies; provide cheap refills and recover the cost from all their other consumers.[35]

Going forward and thinking of the medium to long-term objective of doing away with the use of fossil fuels altogether to get to net zero, gas would need to be replaced by electricity which in turn is gradually going to become less carbon intensive and finally carbon free. Seen from this perspective, gas becomes an intermediate solution. The transition to electricity is easier to promote. State governments have started giving higher quantities of free electricity to households in fulfilment of the election promises of the parties in power ranging from 100 to 300 units per month.[36] In 100 units basic cooking of simple meals can be done. The key challenge is to promote the sales of the newer variants of induction stoves developed in India in recent years to get ordinary households to discover that Indian food can be cooked conveniently with electricity. There is a strong case for doing bulk procurement of the cooking stoves to get substantially lower prices and distribute them through the DISCOMs with the cost being recovered in instalments. This was done successfully in getting households to switchover to energy-efficient LEDs from traditional incandescent light bulbs.[37] The subsidy on LPG needed from the central government would decline. The subsidy for free electricity comes from the state governments. Further, the per unit cost of energy in electricity is lower than the per unit energy cost in LPG. Taking an overall macro view, the subsidy needed for electricity is lower than for LPG.

Biogas is emerging as an option that is worth pursuing now. Cow dung which was being used for cooking is now available in abundance. India has the largest cattle population in the world. Compulsions of electoral politics led to promises by political parties in some states of buying cow dung at Rs 3 per kg.[38] Cow dung is a renewable source of energy. It can be converted at a far lower cost in a decentralized manner at the village level into useful commercial energy. The use of biogas was promoted in a big way in the 1980s. It failed as pilot projects were not done to bring the technology to a level where biogas for cooking could be supplied to households round the clock at full pressure to take care of all cooking needs.[39] The maintenance back up for repair services was not put in place. Then no business model for sustainable operation was put in place. Even if the capital investment is a grant, operation and maintenance have to be financially sustainable. The operator in the village has to be a franchisee who should earn enough to have the right incentive to grow the business.

A new programme, the PM Surya Ghar: Muft Bijli Yojana was launched in 2024 under which 10 million households would be provided with rooftop solar panels.[40] There would be a subsidy of 60% of the cost of the panel with a loan provision for the remaining 40% so that there is no upfront expenditure needed from the beneficiary. Each household would get 300 units of free electricity per month from these panels. Excess generation of electricity during the day would be purchased by the DISCOM. This would help in the repayment of the loan taken for the solar panel. Switchover to electric cooking with this free electricity rather than using LPG would be the rational decision.

Conclusion

India has moved rapidly to provide clean cooking energy to all after taking electricity to all homes. It is also addressing the issue of affordability of clean energy for weaker sections of society.

India can take legitimate pride in what it has achieved.

Annexes

Annexe 1.A.: Performance of Distribution Utilities

The cumulative losses in the distribution sector in India have been rising. These are due to reasons which vary across states. They range from inadequate tariff increases to governance problems. These have been analysed in the TERI discussion paper—*Resolving the Crisis in Power Distribution in India.* [41]

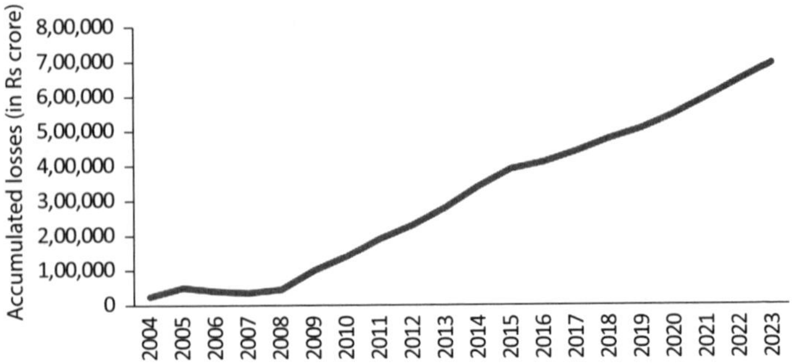

Figure 3 Performance of distribution utilities [41]

Annexe 1.B.: Outlay for Village and Household Electrification

Fiscal year	Outlay (in Rs crore)
2006–07	1,100[#]
2007–08	3,983
2008–09	5,055[&]
2009–10	5,400
2010–11	5,500
2011–12	6,000
2012–13	4,900
2013–14	4,500
2014–15	5,144
2015–16	4,500
2016–17	3,000
2017–18	4,700
2018–19	6,300
2019–20	5,900
2020–21	4,100
2021–22	1,200
Total	**71,282**

Source Ministry of Power Annual Reports and Saubhagya Dashboard

[#] Rural electrification budget. Details available at <https://www.indiabudget.gov.in/budget_archive/ub2006-07/bh/bh1.pdf>

[&] RGGVY subsidy. Details available at <https://www.indiabudget.gov.in/budget_archive/ub2009-10(I)/eb/sbe74.pdf>

Annexe 1.C.: Saubhagya: Pradhan Mantri Sahaj Bijli Har Ghar Yojana [42]

Salient features

- Free metered connection for economically poor households and a charge of Rs 500 (adjusted in electricity bills in 10 monthly instalments) for households other than poor, only after release of connection.
- Establishing camps in villages/ cluster of villages for on spot registration.
- Use of mobile app for identification of beneficiaries and electronic registration including requisite documentation.
- Web-based near real-time monitoring and updating of progress.
- Special-purpose vehicle (SPV)-based stand-alone systems for households located in remote/ difficult areas
- Communication plan for creating awareness about the scheme and its benefits.
- Flexibility to states in mode of implementation (departmental/turnkey/semi-turnkey).

Annexe 1.D.: Salient Features of the Electricity Act, 2003

Preamble

"An Act to consolidate the laws relating to generation, transmission, distribution, trading and use of electricity and generally for taking measures conducive to development of electricity industry, promoting competition therein, protecting interest of consumers and supply of electricity to all areas, rationalization of electricity tariff, ensuring transparent policies regarding subsidies, promotion of efficient and environmentally benign policies, constitution of Central Electricity Authority, Regulatory Commissions and establishment of Appellate Tribunal and for matters connected therewith or incidental thereto."

Salient features

- The law dismantled the monopoly State Electricity Boards created by the Electricity (Supply) Act of 1948 and created separate companies for generation, transmission and distribution.
- It delicensed generation, creating a market.

  "Any generating company may establish, operate, and maintain a generating station without obtaining a licence..." (Electricity Act 2003; Section 7)

  It established an all-India market that encouraged competitive investment in electricity generation.
- It safeguarded consumer interests by mandating that distribution licensees, the distribution companies (DISCOMs), have the universal supply obligation within their licensed areas by stipulating that:

  ".....licensee, shall, on an application by the owner or

occupier of any premises, give supply of electricity to such premises, within one month after receipt of the application requiring such supply" (Section 43).

- Consumer tariffs were regulated by independent State Electricity Regulatory Commissions to ensure fairness and transparency.

  "The state commission shall ... determine the tariff for generation, supply, transmission and wheeling of electricity, wholesale, bulk or retail, as the case may be, within the state" [Section 86 (1) (a)]).

- The state governments could direct the DISCOMs to provide subsidized supply of electricity to categories of consumers such as farmers and low-income households. But then the state governments were required to pay the DISCOMs money for such subsidized supply.

Section 65

"If the state government requires the grant of any subsidy to any consumer or class of consumers in the tariff determined by the state commission.....the state government shall...pay in advance....the amount to compensate..."

Annexe 1.E.: Power Capacity Addition by Public and Private Sectors

Period	Public sector (in MW)	Private sector (in MW)
2007–08	7,335	3,397
2008–09	2,536	2,369
2009–10	5,298	6,135
2010–11	6,481	7,747
2011–12	8,486	17,515
2012–13	9,133	14,583
2013–14	5,906	19,305
2014–15	7,209	19,141
2015–16	10,054	22,320
Total	**62,438**	**90,192**

The data from 2007–08 to 2015–16 shows the fundamental transformation brought about by the Electricity Act, 2003.

In 2007–08, the public sector added 7,335 MW, more than twice the private sector's 3,397 MW. However, by 2009–10, the private sector had begun to close the gap with an addition of 6,135 MW, slightly ahead of the public sector's 5,298 MW.

A dramatic shift occurred starting in 2010–11, when the private sector added 7,747 MW, outpacing public sector additions of 6,481 MW. In 2015–16, the private sector added 22,320 MW, more than double the public sector's 10,054 MW.

Annexe 1.F.: Annual Energy and Peak Demand Growth Rate

The graph illustrates the year-on-year growth rates of energy requirement and peak power demand in India over a 10-year period.

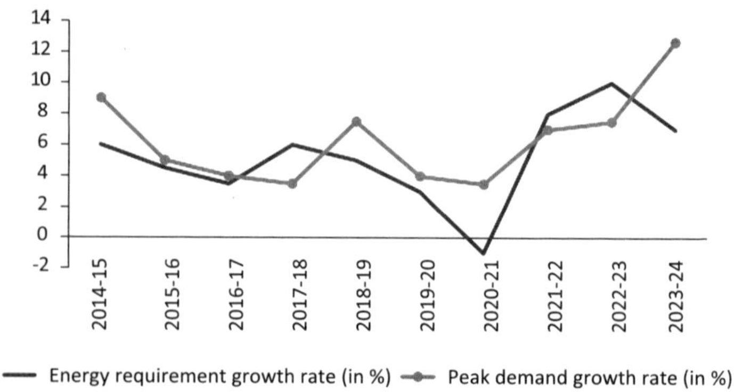

Figure 4 Energy growth rate: requirement and peak demand

From 2014–15 to 2018–19, both energy requirement growth and peak demand growth showed moderate fluctuations, with neither demonstrating a clear upward or downward trend.

A notable dip occurred in 2019–20 and 2020–21, where both indicators dropped sharply, with the energy requirement growth rate turning negative in 2020–21. This decline corresponds with the COVID-19 pandemic, in which a sharp reduction in industrial and commercial activities occurred. The post-pandemic period witnessed a strong recovery.

In 2021–22, energy requirement growth rebounded to 8%, and peak demand growth began to climb steadily. By 2023–24, peak demand growth surged to 12.68%, far more than energy demand growth. This trend of higher peak demand growth is going to persist as air conditioning of workplaces and homes rise.

DISCOMs have therefore to start projecting peak demand growth and make least-cost arrangements for supplying reliable power supply for peak demand periods.

Annexe 1.G. Beneficiaries Under Pradhan Mantri Ujjwala Yojana

Lok Sabha Starred Question No. 59[+]

Answered on 6 February 2025

Excerpts from answer

Pradhan Mantri Ujjwala Yojana (PMUY) was launched in May 2016 with an objective to provide deposit-free LPG connection to adult women from poor households across the country. The target to release 8 crore connections under PMUY was achieved in September 2019. To cover the remaining poor households, Ujjwala 2.0 was launched in August 2021 with a target to release 1 crore additional PMUY connections, which was achieved in January 2022. Subsequently, Government decided to release 60 lakh more LPG connections under Ujjwala 2.0 and the target of 1.60 crore Ujjwala 2.0 connections has been achieved during December 2022. Further, Government approved release of additional 75 lakh connections under PMUY Scheme for the period FY 2023–24 to 2025–26 which has already been achieved during July 2024. As a result of these efforts, LPG coverage in the country has improved from 62% in April 2016 to near saturation now. No targets for new connections under PMUY have been set.

Since launch of PMUY till financial year 2022–23, Government has been bearing an expenditure of up to Rs 1600 per PMUY connection for security deposit (SD) of cylinder, pressure regulator, *suraksha* hose, DGCC booklet, and installation charges. From financial year 2023–24, this support has been increased to Rs 2,200 per connection

[+] India's Energy Transition: power sector vision 2047/2070 and the road map. Presentation made by Ghanshyam Prasad, Chairperson, CEA on 10 January 2025ub2009-10(I)/eb/sbe74.pdf>

for 14.2 kg single-bottle connection/5 kg double-bottle connection and Rs 1,300 per connection for 5 kg single-bottle connection.

To make LPG more affordable to PMUY consumers and ensure sustained usage of LPG by them, government started a targeted subsidy of Rs 200 per 14.2 kg cylinder for up to 12 refills per annum (and proportionately pro-rated for 5 kg connections) to the PMUY consumers in May 2022. In October 2023, Government increased the targeted subsidy Rs 300 per 14.2 kg cylinder for up to 12 refills per annum (and proportionately pro-rated for 5 kg connections). After a targeted subsidy of Rs 300 per cylinder to PMUY consumers, Government of India is providing 14.2 kg LPG cylinders at an effective price of Rs 503 per cylinder (in Delhi). This is available to more than 10.33 crore Ujjwala beneficiaries, across the country.

(in Rs crore)

Particulars	2021–22	2022–23	2023–24
Pradhan Mantri Ujjwala Yojana (PMUY) *	1,618	8,010	8,500

* This includes expenditure on targeted subsidy to PMUY households w.e.f. May 2022.
Source Petroleum Planning and Analysis Cell (PPAC)

In addition to above, during COVID-19 pandemic, as a pro-poor initiative, Government announced a scheme for providing up to 3 (three) free of cost LPG refills to Ujjwala beneficiaries w.e.f. 1 April 2020 under the Pradhan Mantri Garib Kalyan Package (PMGKP). OMCs transferred Rs 9,670.41 crore to bank accounts of PMUY beneficiaries for buying LPG refills and PMUY beneficiaries availed 14.17 crore refills under the Scheme.

As on 1 January 2025, out of 10.33 crore PMUY connections, 3.13 crore PMUY connections are under SC/ST category.

State/UT-wise details of connections released under PMUY Scheme: as on 1 January 2025

State/UT	Number of PMUY connections
Andaman and Nicobar Islands	13,824
Andhra Pradesh	9,73,377
Arunachal Pradesh	53,797
Assam	50,97,770
Bihar	1,16,30,069
Chandigarh	2,027
Chhattisgarh	38,01,853
Dadra & Nagar Haveli and Daman & Diu	17,797
Delhi	2,59,716
Goa	1,955
Gujarat	43,08,874
Haryana	11,15,266
Himachal Pradesh	1,50,787
Jammu and Kashmir	12,69,853
Jharkhand	38,95,672
Karnataka	41,47,264
Kerala	3,87,838
Ladakh	11,086
Lakshadweep	361
Madhya Pradesh	88,47,268
Maharashtra	52,18,129

Contd...

State/UT	Number of PMUY connections
Manipur	2,24,928
Meghalaya	3,17,157
Mizoram	36,006
Nagaland	1,22,150
Odisha	55,50,071
Puducherry	19,381
Punjab	13,59,509
Rajasthan	73,81,916
Sikkim	19,887
Tamil Nadu	41,00,333
Telangana	11,84,145
Tripura	3,16,443
Uttar Pradesh	1,85,95,002
Uttarakhand	5,30,222
West Bengal	1,23,75,962

Source Indian Oil Corporation Limited on industry basis

CHAPTER 2
Breakthrough in Renewables

Introduction

From the beginning of its development journey after gaining independence, India has had the ambition to try and get to the global frontiers in science and technology. It has been allocating scarce resources in this effort for long-term gains.

Energy being the key pillar of the modern industrial economy, India began its programme for the development of nuclear energy in the 1950s.[1] In the early 1970s, following the oil price shock which had a major adverse impact not only on the Indian economy but also on most economies in the world, India began to look for alternative sources of energy. This led to the creation of a new Department of Non-conventional Energy Sources (DNES) in 1982.[2] This grew into a new ministry—Ministry of Non-conventional Energy Sources—in 1992. Forward-looking programmes for the development of new energy sources such as solar, wind, biogas, biomass, geothermal, and ocean energy were taken up. Technology development, pilot projects, and deployment were supported. Even before the new Department was set up, the government had set up Central Electronics Limited in 1974. This enterprise became a pioneer in producing solar photovoltaic (PV) systems. Their PV systems were installed in many remote settlements where the electricity grid seemed unlikely to reach in the foreseeable future.

In 2006 the Ministry of Non-conventional Energy Sources was renamed the Ministry of New and Renewable Energy (MNRE) to better reflect its mandate and work. India had been stepping up its efforts to promote renewable energy to meet the global challenge of climate change. It was acting according to the principle of 'common but differentiated responsibility' that it had successfully advocated, and which was incorporated in the UN Framework Convention on Climate Change (UNFCCC). This was negotiated and signed in 1993 after the global summit in Rio in 1992 which recognized the enormity of the challenge posed by climate change and the necessity of collective action.

Renewable energy, however, was far more expensive than conventional energy. Resources with the government were limited. Claims for more pressing challenges of development were enormous. This naturally constrained the scale of the renewable energy programmes which could be supported by funds from the government. The challenge for successive decision-makers has been to try and find creative ways of overcoming this binding constraint. Finance ministries usually find it difficult to provide more money in the budget. They are, however, a bit more amenable towards granting tax concessions. The case for tax concessions becomes stronger if they are given for new investments which need to be promoted but are unlikely to take place without the tax concessions. In these cases, no revenue is being generated and arguably there is, therefore, no loss of revenue.

Notwithstanding such severe financial constraints, India's achievements in renewables have been far more than expected. This has surprised many. India now has the third-largest renewable energy capacity in the world—105.65 gigawatt (GW) of solar and 50 GW of wind power as on 31 March 2025.[3] Renewable energy capacity has been growing at a compound annual growth rate (CAGR) of 9.94%. [4] This capacity has been primarily created by private investment. This was the result of government policies, creating the right incentives for profit maximizing private firms to invest in renewables. Getting private firms to invest for achieving desirable outcomes is never easy. This is the case not only in India but also in the rest of the world. Success has been the result of innovative smart policies. In this chapter, this extraordinary success story is described and the policies that made this possible are analyzed.

Wind Power

Wind power was a mature proven technology in the last century. It has been the main pillar of the growth of renewable energy in the advanced economies, especially in the countries where there is abundant wind power potential. For instance, Germany got around 29% of its electricity from wind power in 2024. The share in the United Kingdom is similar, around 28%. [5]

India first achieved success in developing renewable energy on scale with wind power. This began in the early 1990s and since then growth has been remarkable. Wind power has been developed in India mainly by the private sector. This was new for India. India's energy sector had been developed almost entirely in the public sector. With rapid growth, India was adding an average of roughly 150 megawatt

(MW) of wind power capacity annually by 2000. With such a large market, a globally competitive wind power-manufacturing industry came up. The industry emerged to meet the growing demand in the domestic market. Figure 2 shows the growth of wind power capacity in the country and Figure 3 shows the private investment that went into creating this capacity (Annexe 2.A.).

Enabling New Policies

Two new policy decisions, one by the central government and the other by state governments, led to the surge in private development of wind power. These were:

1. The fiscal incentive of accelerated depreciation provided by the central government.
2. The provision of the innovative 'banking' mechanism by the state governments.

These made private investment in captive wind farms very profitable. Investments and capacities grew in the states where there was good wind power potential.

Accelerated Depreciation

The Ministry of Finance was persuaded to grant accelerated depreciation of 100% against the prescribed rate of 25% in 1994 for wind energy projects.[6] This was a special dispensation to promote wind power development in the country. This gave sizeable tax benefits to the investors in wind power. Private sector investment in captive wind farms took off. These investors harnessed wind energy to meet their demand for electricity. At the same time, they were able to reduce their tax liability from their other profitable businesses substantially by availing 100% depreciation for their investment in captive wind farms.

Private investments created the domestic market for wind turbines, which in turn spurred to the growth of the manufacturing industry and other allied services. During the initial years, wind capacity was developed primarily for captive consumption. From 2004 to 2014, the wind sector grew at a phenomenal compound growth rate of 24%.

This provision of accelerated depreciation was withdrawn in 2012 as the Finance Ministry felt that wind power development had gained sufficient momentum, and this special promotional dispensation was no longer required. The Ministry of New and Renewable Energy was able to persuade the Finance Ministry to reinstate accelerated depreciation in 2014 to maintain the momentum of wind power development which had declined after the 100% depreciation dispensation had been withdrawn. In 2017, the Finance Ministry's view did prevail to a considerable extent. The rate of depreciation was lowered to 40% and since then it has stayed there.

This experience of wind power development illustrates the enormous success in inducing private investment in a particular desired direction if the risk-free returns appear attractive to private investors. Substantive tax concessions can be a very powerful and effective instrument in certain circumstances.

Banking of Wind Power

The concept of banking was first introduced in the state of Tamil Nadu and since then it has been utilized by several other states.[7] Under 'banking' the State Electricity Board took the electricity generated by a captive wind farm of an industrial enterprise and banked it. This banking was notional. The industrial enterprise which receives electricity from the DISCOM was billed on a net basis after deducting the quantum of wind power it had banked. The banking mechanism was profitable for industrial consumers. The tariff for industrial consumers ranged between Rs 7 and 9 per kilowatt hour (kWh). The banking provisions allowed them to use electricity at a cost of Rs 4–5/kWh, the actual cost of electricity generation from their captive wind farm. This resulted in large savings.

This effective policy instrument was available as India has the special characteristic of having a cross-subsidy regime in its electricity supply rates. The high-end industrial and commercial consumers are charged more than the cost of supply. This extra revenue is used to supply electricity below cost to consumers of the weaker sections. Cross-subsidy is a de facto tax within a sector where wealthier consumers pay to subsidize supply to consumers from the weaker sections. Under 'banking', the State Electricity Boards did not get the extra revenue from the subsidizing industrial consumer to the extent he generated electricity from his captive wind farm. The state government took this decision to attract private investment in the development of wind farms with this policy. Given the volume of their total turnover, this marginal loss of cross-subsidizing revenue from subsidizing industrial consumers was affordable. The financial gain from captive wind power became an added incentive for industrial units to invest in captive wind farms.

To increase the pace of wind power development, tariff-based competitive bidding as was being done for solar power projects was started in 2017 for wind power projects by the Ministry of New and Renewable Energy.[8] This was for the supply of wind power to the DISCOMs through a long-term power purchase agreement (PPA).

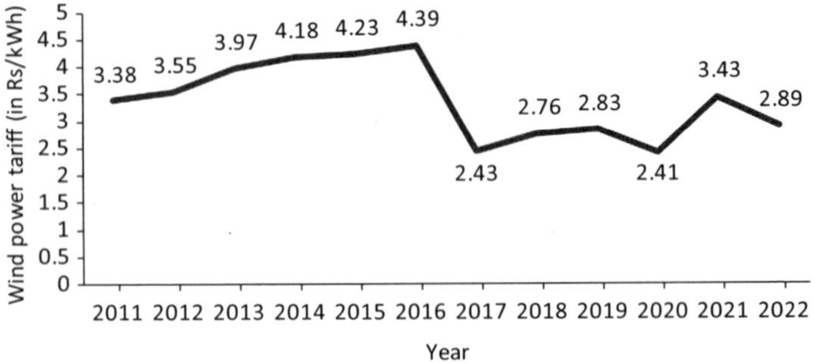

Figure 1 Trend in wind power tariffs in India

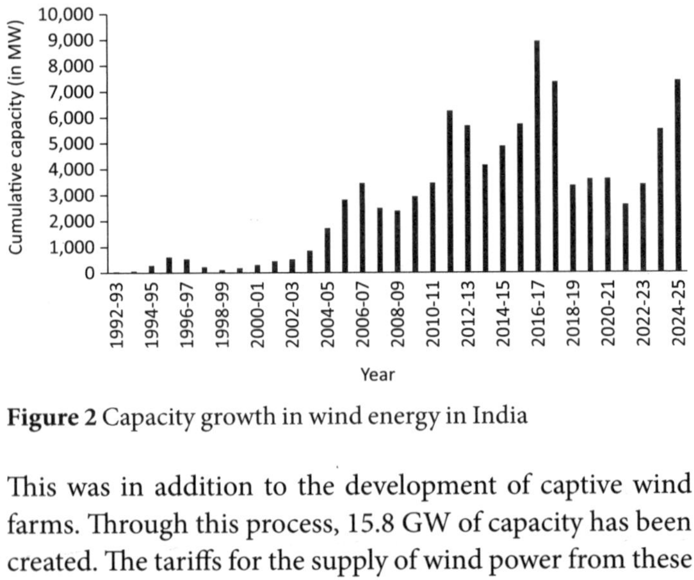

Figure 2 Capacity growth in wind energy in India

This was in addition to the development of captive wind farms. Through this process, 15.8 GW of capacity has been created. The tariffs for the supply of wind power from these bids have been reasonable and attractive for the DISCOMs.

However, the wind power potential in India has constraints which are distinct. There are limited states which are suitable for harnessing wind power. These states are Andhra Pradesh, Tamil Nadu, Telangana, Gujarat, Karnataka, Maharashtra, Madhya Pradesh, and Rajasthan

(Annexes 2.B. and 2.C.).[9] Further, wind in these states does not blow steadily. There are large variations in the day and over the seasons, especially during the monsoon. Winds gain speed in India's monsoon season which lasts from June to September. Two-thirds of the total wind energy is produced during the monsoon season.[10] This seasonality of wind power and its potential being restricted to only a few states limits the extent of feasible wind power development in India. Solar power development has now overtaken wind power capacity creation.

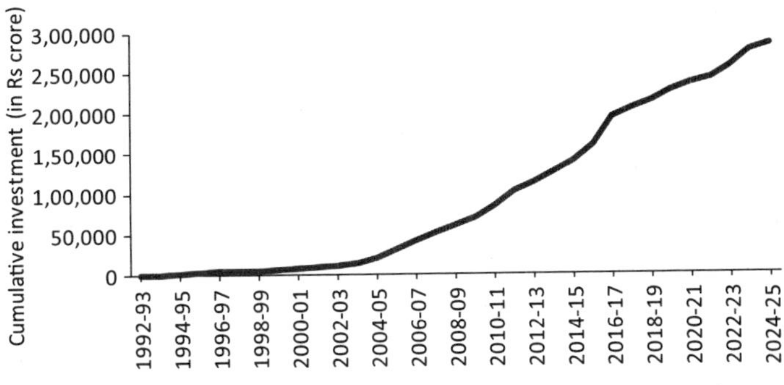

Figure 3 Capacity investment trends in wind energy in India

Mandatory Promotion of Renewable Energy

Around 2000, India felt that it needed to do more to increase the share of renewables in its energy mix. At that time, India had achieved some success with wind power development. Other than wind power nothing else was gathering momentum as the costs were much higher than the cost of electricity from conventional sources.

The idea that an increase in the share of renewables be mandated by law emerged and started receiving serious attention. This, however, presented the challenges listed here.

- India was a large diverse country with many states. Electricity was initially a 'state subject'. The supply of electricity was being undertaken by the State Electricity Boards. Electricity had been made a 'concurrent subject' in the 1970s through a Constitutional amendment giving the central government legislative power. India had a coalition government then at the Centre and regional parties and the states were being more assertive. Would a central law mandating a percentage of electricity supply to come from expensive renewables be acceptable to the states? Even if enacted, would it be implementable?
- What percentage should be prescribed under the law and on what basis should this be determined?
- The natural endowments for renewable power generation and their costs varied considerably across states.
- As renewable energy was more expensive, how would the higher cost be absorbed? This was an obstacle even in the advanced economies. In India, this presented a major, almost insurmountable, challenge. Many of the State Electricity Boards were in serious financial distress with accumulating losses resulting from revenues being less than costs. The central government in 2001 had for the first time put together a bailout package for the State Electricity Boards.
- Legislation in India is a lengthy process. Increasing the share of renewables in the future would become a slow tedious process.

After considerable deliberation, the idea was dropped, and another solution emerged.

The Electricity Act, 2003

A new law for electricity replacing the existing three laws was being evolved. It was going to restructure the power sector by doing away with the State Electricity Boards and creating separate companies for generation, transmission, and distribution. In addition, it would create independent State Electricity Regulatory Commissions (SERCs) to regulate the electricity sector in the states.

An explicit provision giving the State Electricity Regulatory Commissions the responsibility to promote renewables and prescribe their share was introduced in the law.

Section 86 (e) of the Electricity Act, 2003 states that the State Electricity Regulatory Commissions must among other functions:

> "Promote co-generation and generation of electricity from renewable sources of energy by providing suitable measures for connectivity with the grid and sale of electricity to any person, and also specify, for purchase of electricity from such sources, a percentage of the total consumption of electricity in the area of a distribution licensee." [11]

The Electricity Act, 2003 gave the State Electricity Regulatory Commissions the responsibility of promoting renewable energy and prescribing the share of renewable energy that a DISCOM had to have in its total electricity supply. This was acceptable to the states as decisions for each state would be taken by their State Electricity Regulatory Commissions after public stakeholder consultations. Fortunately, there was a broad consensus among political parties and the states that India should promote renewables and increase their share even if these were more expensive. With agreement in principle, the issue became one of the rate of growth of renewables to be prescribed in a state considering affordability and sustainability.

The regulatory framework for trading in Renewable Energy Certificates (RECs) was created by the Central Electricity Regulatory Commission in 2010.[12] This was a forward-looking measure which considered the fact that renewable energy resources were unevenly spread across the states of India. The Renewable Energy Certificates emerged as a tradeable commodity. The DISCOM in a state where renewable energy resources were low could buy the Renewable Energy Certificates in the market and meet its mandated renewable energy purchase obligation. Other DISCOMs could convert their renewable energy contracts into the Renewable Energy Certificates and sell them in the market at a premium.

Solar Energy

Though holding great promise, solar energy remained considerably more expensive than conventional thermal power for a long time. This prevented large-scale deployment the world over. India was no exception.

In 1993, a scheme was started to install 50,000 solar pump sets for irrigation. Solar power for pumping required high upfront expenditures as solar panels were quite expensive at that time. As subsidy provisions were limited, the achievements were modest.[13] By 2015, around 18 thousand solar pumps had been installed. In 2001, the Ministry of Non-conventional Energy Sources initiated the Village Electrification Programme for unelectrified villages and hamlets with the installation of solar PV systems.[14] Ninety per cent of the cost of these was to be borne by the central government as a grant. This programme, however, did not take off. There was no capacity in the country to put up village-level installations on a large scale. The programme was not well promoted and communicated among the people in unelectrified rural areas. Claiming and getting subsidies for small village-level projects is at the best of times difficult in the vast underdeveloped country that India is.

The scale of deployment of solar power remained modest and was only 3 MW till the launch of the National Solar Mission in 2010.[15]

National Solar Mission[16]

In a major act of responsible climate leadership, Prime Minister Manmohan Singh decided in the summer of 2009 that India would create 20 GW of solar power capacity by 2022. He tasked his team, which was taken by surprise by the boldness of the ambition, to design the programme for achieving this goal. This led to the launch of Jawaharlal Nehru National Solar Mission (JNNSM) in 2010.

The government at this stage was trying to rein in the fiscal deficit. It was in no position to provide large sums of money for this new programme. The design of the Mission posed seemingly insurmountable challenges. Would it remain another one of those big announcements with modest and disappointing results?

Fortunately, the Mission turned out to be a surprising success, far more than the most optimistic assessments at that time. The Solar Mission was creative in its implementation strategy with many features being adopted for the first time in India. A major departure from past practice and precedents took place. There were innovations in the programme design.

- The central government in departure from traditional practice decided that instead of offering financial support to private investors or the state governments for developing solar power it would assume direct responsibility for achieving the highly ambitious target of the Mission. It recognized that solar power would have to be purchased by DISCOMs and supplied to their consumers.
- The process of inviting tariff-based bids for grid-connected solar power projects was adopted. This marked a break with prevailing practice. The practice in the Ministry of New and Renewable Energy had been to determine the cost of the product being promoted. The subsidy needed to get the new, more expensive green product buyers in the market would emerge. This would then be given to the producer or the buyer. The money budgeted for the subsidy would determine the maximum volume whose production and sale could take place in the market.

When the decision to invite bids was being taken, there were apprehensions that bids may not be forthcoming. The programme would then be a non-starter which would be embarrassing. After all, this had not been the practice adopted elsewhere.

The successful programmes in other countries had adopted the feed-in-tariff approach. Under this a remunerative price at which solar power would be bought by the power utility was announced and then the market responded with investments in solar power. Government paid the utilities the difference between the higher cost of solar power and conventional power. Germany implemented a large solar power programme with a feed-in-tariff. This created the volumes of the global solar power panel market which then led to economies of scale and innovation in production. This was followed by a steep reduction in costs and prices.

Then there was the risk that someone may bid aggressively, offer a non-viable price and might not implement the project. Prior experience and capability were not made prequalifying conditions, as the programme in its initial phase needed to keep the bidding process non-restrictive to get new entrants and competition. The counter view which prevailed was that it would be the successful bidder's loss if he was not able to implement the project. There would be no financial loss to the DISCOM as money would have to be given only for the solar power supplied. Failure to supply would have no adverse consequences for the DISCOM as the quantity being contracted would be marginal to the total purchases of the DISCOM. The bid strategy:

- A series of bids were invited starting with small quantities. This had the benefit of reducing the funds needed by prospective bidders as well as their risks.
- Then, in a major departure from normal practice the entire quantity for which bids were invited was not contracted to the bidder who had quoted the lowest price. Projects were awarded to a few bidders in each round to create a new competitive industry structure of solar power developers (Annexe 2.D.). Once the supply side of the market was well developed, this practice was discontinued.

- Further de-risking was provided to the developers by getting them to enter into power purchase agreements for the supply of solar power to a few DISCOMs, reducing their risk from having only one buyer.
- The bidders had the freedom to choose the location of their plants. They had to supply power to an existing 33 kV substation.
- Solar power was provided zero-cost transmission through the integrated national all-India grid, reducing the cost of solar power for all DISCOMs.
- As the programme gathered momentum, Solar Parks were developed by the states with financial assistance from the Ministry of New and Renewable Energy where land was made readily available for solar power projects. Then, as project sizes grew and many projects were taken up in contiguous areas where land was readily available with high solar radiation, green transmission corridor development was taken up to ensure that transmission did not become a constraint in the evacuation of solar power.

Bundling Mechanism

When the Solar Mission was started, solar power was far more expensive than new thermal power. Why would state DISCOMs buy such expensive solar power? The central government just did not have the money to subsidize solar power to make its price the same as that of thermal power for a large programme.

This seemingly insurmountable problem was resolved through the innovative process of bundling. Bundling was the mechanism used to launch the National Solar Mission without any budgetary support from the government.

Bundling enabled the sale of solar power even though its cost was much higher than new thermal power. This was feasible due to the specific features of the Indian power supply system where tariffs for power generation were regulated. Tariffs were determined for each power plant and each unit separately. Under the bundling mechanism, costly solar power was bundled with relatively cheaper power from old, depreciated coal-based thermal power plants of NTPC, a central government public sector undertaking. Fifteen per cent of the capacity of each thermal plant of NTPC was available to the central government for discretionary allocation. This system was in place from the outset when NTPC was set up. Its plants were regional, and power was allocated to the states according to a settled formula. Fifteen per cent was kept as unallocated power to enable the Centre to provide some additional power to a state facing greater need. The reasons for this could vary; additional demand in a state for irrigation, or a state having failures in its own generating stations, resulting in shortages. The regulated Indian power supply system had tariffs determined for each plant to give a reasonable return on investment. As a thermal plant became old, there was no debt to be repaid and return on equity was on face value. Inflation kept reducing its value in real terms. The tariff of power from an old plant was very cheap in comparison to any new plant as the fixed cost had become nominal. To illustrate, bundling was done with Units 1 and 2 of NTPC's Singrauli plant, where the tariff was 1.81 Rs/kWh in 2010–11.[19]

The cost of the bundled power was made comparable to conventional thermal power from new projects by combining one unit of expensive solar power with four units of cheap thermal power from old NTPC plants.

DISCOMs were therefore willing to buy this power through normal power purchase agreements. This enabled the Solar Mission to take off without any financial support from the government. Phase 1 of the Mission was implemented through the power trading subsidiary of NTPC making the bundling of cheaper NTPC power with solar power a smooth, seamless operation.

Viability Gap Funding

After the success in the initial phase in getting attractive bids from private solar power developers, the programme needed to be scaled up. The availability of unallocated power from NTPC was limited.

Viability Gap Funding (VGF) was then introduced in 2013 to take the programme to the next level under Phase 2 of the Solar Mission. [20] The implementation of the programme was now entrusted to the Solar Energy Corporation of India (SECI). The Solar Energy Corporation of India had been set up by the Ministry of New and Renewable Energy. The viability gap funding was a capital subsidy to bridge the gap between the project cost and the cost needed to provide electricity at the prevailing rate for new thermal power. At this rate, the DISCOMs would buy solar power as their power purchase costs remained the same. For a given tariff, bids were invited with the viability gap funding being the bid parameter. So, whoever asked for the lowest viability gap funding would get the project. The tariff to be paid to the developer was pre-fixed by Solar Energy Corporation of India and was part of the bid. Under this scheme, over 5,000 MW of solar power projects were developed.[21] The tariffs continued to fall rapidly. Then bids were received where no viability gap funding was asked for. This became the major inflection point for solar power development in

India. Financial support was no longer needed for solar power to be able to get purchasers. This scheme of viability gap funding became redundant and was discontinued in 2017. Since then, solar projects have been coming up through competitive bids. Prices have kept declining. The price of solar power has become considerably lower than the price of new thermal power.

Many state DISCOMs began to invite bids on their own and entered into contracts for the supply of solar power. This gave added momentum to solar power development in the country.

India was able to get the full benefit from the global decline in production costs of solar panels through its policy of procuring solar power by inviting bids for long-term supply under standard power purchase agreements. A strong competitive private industry structure emerged. There has been innovation and cost reduction in the deployment and maintenance of solar power. As a result, India has amongst the lowest rates of solar power in the world.

Many developers have grown into large firms. They have been able to raise debt and equity from international financial markets. Some have even got listed in foreign stock exchanges. To illustrate, ReNew Power is listed on NASDAQ. Finances have not been a constraint for the growth of renewable energy in India. And this growth has been accelerating (Annexe 2.E.).

The growth of solar power in India under the Solar Mission was very rapid. And so was the decline in costs. The competitive bid process resulted in prices falling far more than initially expected. When the Solar Mission commenced in 2010–11, the solar tariff approved by the

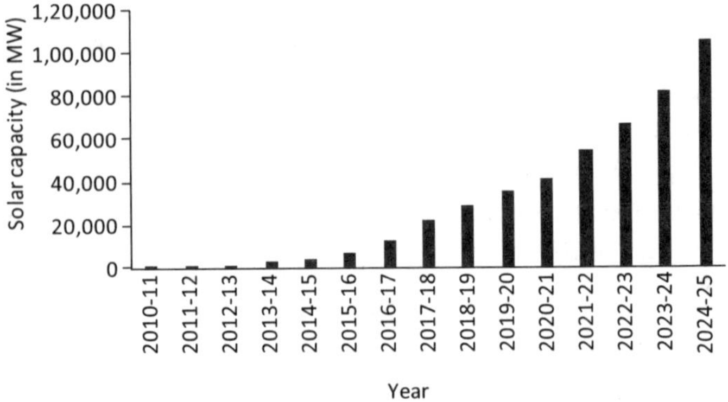

Figure 4 Solar power capacity creation in India

Central Electricity Regulatory Commission was Rs 17.91/kWh. The Mission was successful in getting the price of solar power to go down substantially through competition. Solar tariffs declined rapidly from around Rs 12 when the Mission began [17] to less than Rs 2.50 now.[18] The remarkable success of China in lowering the cost of production of solar panels was the key to the decline in prices. Developers bought their panels from China.

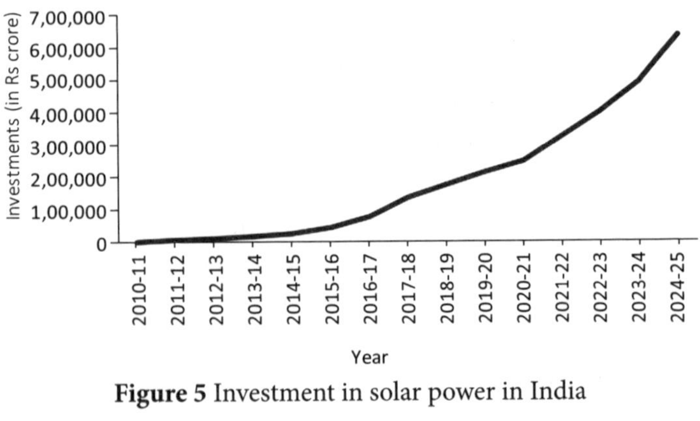

Figure 5 Investment in solar power in India

Figure 6 shows the decreasing trend of solar tariffs in India. From being more expensive than thermal power, solar power is now decisively cheaper.

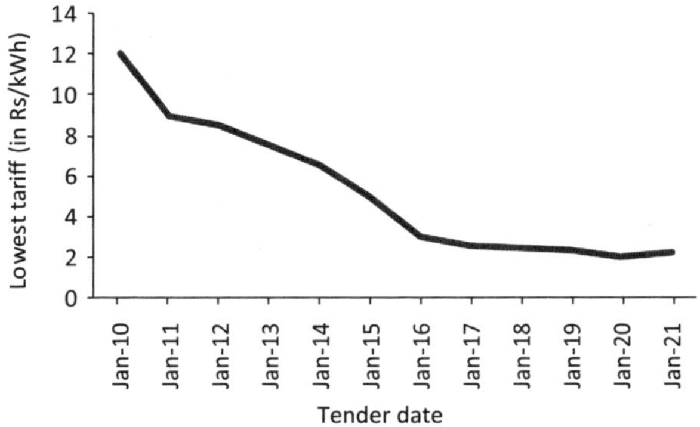

Figure 6 Decline of solar prices in India
Source MNRE (2021)

Key Learnings

Looking from an investor's perspective in taking decisions on potential investment opportunities is a good starting point in trying to understand India's experience. First, the investor must see reasonable returns on his investment. Investments are made only when there is an expectation of profits. Then the investor makes his assessment of risk in the potential investment. The expected return and risk calculus determines the decision of a particular investor. Some will take greater risk provided the expected returns are higher. Others would only invest if the risk perception is low, and would be satisfied with lower returns.

In a competitive bid process, all the bidders expect to get returns that they are comfortable with in case they get the contract at their bid price. In the bid process followed for solar power, the purchase of the entire power production was assured for 25 years through the long-term power purchase agreement. The successful bidder was assured risk-free returns. There was neither a price risk nor an offtake risk. This made the investment practically risk-free. The power purchase agreements were with the state DISCOMs. Being state-owned utilities providing such an essential service, there was an implicit state guarantee behind these power purchase agreements. These state DISCOMs may have losses and financial difficulties, but at the end of the day they are bailed out and kept afloat as they cannot be allowed to go under. Hence, the investor's response to these bids has been good and competition has nudged prices downwards.

Such a risk-free investment made it easy for the successful bidders to raise finance, both for equity and debt, from domestic as well as international markets. Debt was readily available and at competitive rates of interest. As a result, Rs 7,44,763 crore of investment in solar power development has been financed by domestic and international financial markets. India's financial markets and international financial markets are interconnected. Funds are readily available for what are assessed by the market as profitable projects. Finance has not been a constraint for these massive investments in renewable energy projects.

It is important to recognize that this success has been possible by the de-risking of competitive private investment through long-term power purchase agreements by DISCOMs. This unique market design has been critical for the success of the programme. The centrality of the DISCOM undertaking assessment of future demand and entering into long-term power purchase contracts for providing reliable supply has to be recognized and appreciated. Without the DISCOMs' power purchase agreements, private investment on this scale would not have been feasible. Many 'reformers' enamoured of full deregulation and 'spot markets' pioneered by the UK in the early 1990s need to rethink and derive the correct lessons from India's situation and experience.

The key lesson, relevant across sectors, is that it is a good idea to begin large ambitious new programmes on a modest scale, learn and adjust the design if necessary, succeed and then scale up. Rapid scaling up on solid foundations of initial learning by all stakeholders gives better results. India's spectacular increase in solar power capacity creation is a good example of this approach.

Greater Confidence and Ambition

The share of solar and wind power in electricity generation began to rise steadily without financial support from the government. This gave India confidence.

India constructively facilitated the Paris Agreement in 2015. Its Nationally Determined Contributions (NDCs) were ambitious. It undertook to achieve a 40% share for renewables in its total installed power-generation capacity by 2030. This was achieved in 2023, well ahead of schedule. [22]

At COP 26, in Glasgow in November 2021, Prime Minister, Narendra Modi surprised everyone when he announced that India would have 500 GW of fossil fuel-free capacity by 2030 when India's total generating capacity then was only 383 GW in 2021.

His announcement of the five nectar elements or the *Panchamrit* at Glasgow was as follows:

> "In the midst of this global brainstorming on climate change, on behalf of India, I would like

to present five nectar elements, *Panchamrit*, to deal with this challenge.

First: India will reach its non-fossil energy capacity to 500 GW by 2030.

Second: India will meet 50% of its energy requirements from renewable energy by 2030.

Third: India will reduce the total projected carbon emissions by one billion tonnes from now onwards till 2030.

Fourth: By 2030, India will reduce the carbon intensity of its economy by less than 45%.

Fifth: By the year 2070, India will achieve the target of net zero. These *Panchamrits* will be an unprecedented contribution of India to climate action.

Today India is moving forward on the subject of climate with great courage and great ambition. India also understands the suffering of all other developing countries, shares them, and will continue to express their expectations.

For many developing countries, climate change is looming large over their existence. We have to take big steps today to save the world. This is the need of the hour and this will also prove the relevance of this forum. I am confident that the decisions taken in Glasgow will save the future of our future generations, giving them the gift of a secure and prosperous life."

A new confident India displaying responsible climate leadership had emerged.

Annexes

Annexe 2.A.: Year-wise Growth Wind Power Capacity Since 1990 [23, 24]

Years	Annual Capacity Addition per year (in MW)	Cumulative Capacity (in MW)	Annual Investment (in INR crore)	Cumulative Investment (in INR crore)
1992-93	15	16	90	96
1993-94	59	75	356	452
1994-95	266	341	1,594	2,046
1995-96	359	700	2,157	4,203
1996-97	155	855	929	5,132
1997-98	62	917	371	5,504
1998-99	45	962	267	5,771
1999-2000	132	1,094	794	6,565
2000-01	182	1,276	1,091	7,656
2001-02	251	1,527	1,505	9,161
2002-03	237	1,764	1,422	10,583
2003-04	669	2,433	4,013	14,596
2004-05	1,120	3,553	6,722	21,318
2005-06	1,751	5,304	10,507	31,825
2006-07	1,812	7,116	10,869	42,695
2007-08	1,595	8,711	9,571	52,266
2008-09	1,473	10,183	8,835	61,101
2009-10	1,607	11,791	9,644	70,745
2010-11	2,341	14,132	14,045	84,790
2011-12	3,081	17,212	18,483	103,274
2012-13	1,803	19,015	10,817	114,090
2013-14	2,027	21,042	12,159	126,249
2014-15	2,326	23,368	13,956	140,206

Contd...

Years	Annual Capacity Addition per year (in MW)	Cumulative Capacity (in MW)	Annual Investment (in INR crore)	Cumulative Investment (in INR crore)
2015-16	3,410	26,777	20,459	160,664
2016-17	5,502	32,280	33,014	193,679
2017-18	1,865	34,145	11,191	204,870
2018-19	1,481	35,626	8,886	213,756
2019-20	2,068	37,694	12,407	226,163
2020-21	1,553	39,247	9,320	235,482
2021-22	1,111	40,358	6,663	242,146
2022-23	2,276	42,633	13,653	255,799
2023-24	3,253	45,887	19,520	275,319
2024-25 (As of September 2025)	1,476	47,363	8,856	284,175

Annexe 2.B.: Wind Power Potential in India[25]

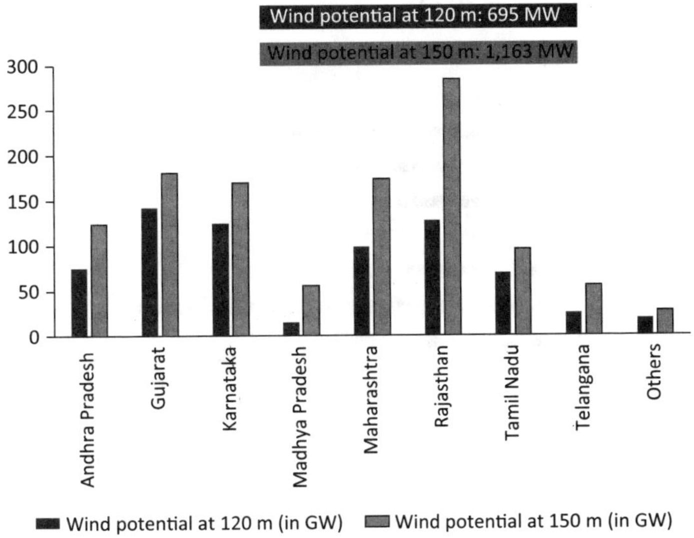

Annexe 2.C.: Installed Capacity of Wind Power in Indian States in 2024[26]

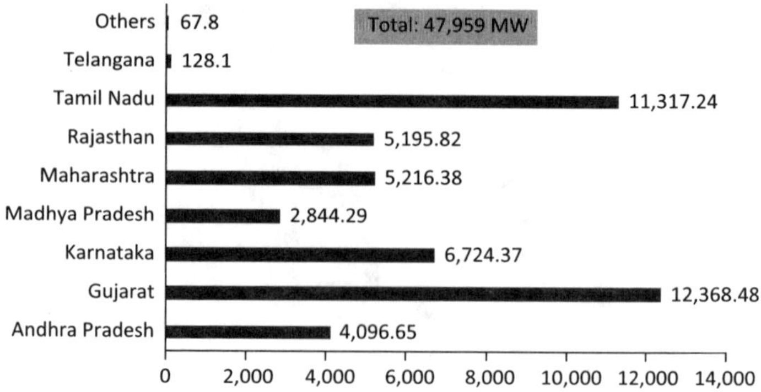

Annexe 2.D.: Initial phase 1 of Solar Power Project Bidding with Bundling under National Solar Mission

Phase I, Batch I

Contracts were awarded for 5 MW each to 30 bidders and 150 MW of solar capacity was created. The average tariff for Batch 1 was Rs 12.12. With this a competitive industry structure was created.

Phase I, Batch II

In Batch II a total capacity of 340 MW was installed across 27 projects and the project size was increased to 15 to 20 MW each. The average tariff came down by over one-fourth to Rs 8.77.

Annexe 2.E.: Year-wise Growth Solar Power Capacity[27]

Year	Annual capacity addition (in MW)	Cumulative capacity (in MW)	Annual investment (in Rs crore)	Cumulative investment (in Rs crore)
2010–11	32.39	32.39	194.34	194.34
2011–12	908.92	941.31	5,453.52	5,647.86
2012–13	745.13	1,686.44	4,470.78	10,118.64
2013–14	1,135.47	2,821.91	6,812.82	16,931.46
2014–15	1,171.62	3,993.53	7,029.72	23,961.18
2015–16	3,130.36	7,123.89	18,782.16	42,743.34
2016–17	5,658.63	12,782.52	33,951.78	76,695.12
2017–18	9,563.69	22346.21	57,382.14	1,34,077.3
2018–19	6,750.97	29097.18	40,505.82	1,74,583.1
2019–20	6,510.06	35607.24	39,060.36	213643.4
2020–21	5,628.78	41236.02	33,772.68	2,47,416.1
2021–22	12,760.52	53996.54	76,563.12	3,23,979.2
2022–23	12783.82	66780.36	76,702.92	4,00,682.2
2023–24	15,033.24	81813.6	90199.44	4,90,881.6
2024–25	23,832.89	105646.5	1,42,997.3	6,33,878.9

CHAPTER 3

The Imperative of Storage

Introduction

Solar energy can be converted into electricity when the sun shines. Wind power is generated when the wind blows. Solar and wind power generation is inflexible, variable, and intermittent. But electricity is needed even when the sun does not shine, or the wind does not blow. The demand for electricity varies considerably during the day and across seasons. This variable demand must be matched by supply at every moment. This is a technical necessity for the functioning of the electricity supply system and the stability of the grid. If the objective is to supply all electricity from renewable energy sources which are intrinsically variable and inflexible, then renewable energy must be stored and supplied to ensure that varying demand is fully met, every moment of the day, and on all days through the year. This is only possible by storing electricity. Stored electricity would enable power supply systems to use only renewables and continue to provide reliable quality power supply to meet variable demand. A zero-emission electricity system can become a reality only with the provision of electricity storage.

DISCOM Perspective

In India solar and wind power have become considerably cheaper than electricity from a new thermal power plant.

This has been the case for some years now. But the true cost of renewable energy must be seen from the perspective of a distribution company (DISCOM), which has to supply reliable electricity to meet the full demand of all its consumers at every moment. As long as the share of renewable electricity was marginal, it did not matter to the DISCOMs and they could continue with business as usual. They provided reliable supply through their long-term power purchase agreements (PPAs) for the supply of electricity from thermal power plants. The purchase price in these contracts for thermal power has two components:

1) The fixed cost which covers the capital cost of the investment in the plant along with operation and maintenance cost.
2) The variable cost which covers the cost of fuel.

The fixed cost per unit of electricity declines as the plant is run at fuller load as it is spread over a larger number of units, and it goes up as the plant is required to produce less. The tariff per unit is the least when the plant load factor (PLF) is the highest and rises as the plant load factor declines. As the share of renewables rises, the plant load factor of thermal plants comes down as they must reduce generation when renewable power is coming into the system. As a result, the DISCOM pays a higher price per unit of thermal power as its fixed cost is spread over fewer units. This higher cost, therefore, needs to be factored in to compute the real cost of renewables to the DISCOM as the share of renewables is set to rise rapidly in the coming years.

But if electricity from contracted renewables and thermal power at full plant load factor were stored when generation was more than demand, then this stored electricity would be available to meet peak demand at other times of the day. The higher plant load factor of thermal plants would keep the unit cost of thermal power at the lowest. Creating storage to absorb all the contracted electricity and then using stored electricity for additional supply to meet peak demand would then become the least-cost option.

Going forward, storage of renewable energy would enable the DISCOM to meet demand outside the hours when renewable energy is generated. It need not enter into long-term power purchase agreements for new thermal power to ensure reliable round-the-clock supply. To meet anticipated demand, the quantum of renewable power generation and storage capacities that would be needed can be created. Some spare storage capacity would also be needed to cater to sudden fluctuations in renewable power generation as well as unforeseen surges in demand.

Even without renewables coming into the picture, the DISCOMs had to fully meet demand, which varied during the day with some hours of peak demand and others of low demand. To meet this varying demand, the DISCOMs were contracting for thermal capacity where the plant load factor would be lower but was needed to meet demand peaks. The implication was a higher per unit cost. In recent decades, the DISCOMs have not tried to make separate arrangements for the supply of peaking power though they should have been doing so. Earlier, there was no need for it as power cuts were normal due to inadequate generating capacity in the country. But this is no longer the case. DISCOMs need to make the least-cost arrangements for meeting peaks in demand. Contracting for thermal capacity to meet only peaking power needs is unnecessarily expensive. Storage of thermal power to meet peak demand is a cheaper option.

Initiation of Storage

A consensus has emerged in the last few years in India that large grid-scale electricity storage capacities are needed. The pace of electricity transition towards renewables is gathering greater momentum. The share of inflexible and variable solar and wind power has already exceeded 10%.[1] The goal of creating 500 gigawatts (GW) of non-fossil fuel capacity by 2030 is being pursued. The creation of adequate storage has to now become an intrinsic part of the implementation strategy with quantified targets. This alone would enable the full utilization of renewable and thermal power generation capacities.

Hence, India needs to move rapidly to create storage. It needs to do it far earlier than the advanced industrial economies as its renewable generation has a predominant share of solar power. Solar power is generated only when the sun shines. This is in sharp contrast to Northern Europe where wind power is the primary source of renewable energy. Wind blows throughout the year and round the clock. Hence, the creation of storage has started gaining attention and priority in these countries only in the last few years.

Germany is aiming to get 80% of its electricity from renewables by 2030.[2] The share went over 60% last year.[3] The creation of storage has therefore been gaining momentum. The need for storage was dramatically highlighted a few years back for the first time. On a day when demand was low in their summer and renewable generation was going to exceed demand, consumers were offered money to consume more electricity, something unimaginable till then. The other options for maintaining grid stability were more expensive. It is now recognized that the goal of carbon-free electricity systems substituting thermal power fully with renewables cannot be achieved without large-scale storage.

The Central Electricity Authority (CEA) of India saw the critical need for creating storage and estimated that India would need 73.9 GW of storage by 2032.[4] In 2022, the Ministry of Power extended the ambit of the renewable purchase obligation (RPO) under the Electricity Act to include storage. [5]

Storage Options

At present, the feasible storage options are as follows:

- Electricity has been stored in batteries for a long time. Batteries have been in use for a variety of needs as the modern industrial economy has evolved; from automobiles to torches to mobile phones and laptops. The use of batteries for grid-scale storage is now taking place.
- Pump storage projects (PSPs) were set up for large-scale grid storage to enable the grid to meet spikes in the demand for electricity.
- Concentrated solar thermal storage is a relatively new technology which has become mature.
- Energy is stored in hydrogen in the same way as it is stored in fossil fuels.

Pump Storage Projects

Pump storage projects[6] can be developed at existing hydropower project sites. They generate electricity in the same way as hydropower is generated; water falls from the reservoir of a dam and the energy of the falling water is used to rotate a turbine to generate electricity. Storage is achieved by using electricity to pump water up the reservoir of the existing dam. For this, a small reservoir is created at the bottom of the dam where water is retained. This water is pumped up using electricity. This stored energy is converted into electricity whenever needed by letting the water come down from the upper reservoir. The energy loss in storage ranges between 20% and 30%. Like hydropower projects, pump storage projects have a long life and nominal operation and maintenance costs.

This is a mature and proven technology of the last century. It was developed to provide peaking power needs, and grid stability in conventional electricity supply systems. Pumped storage hydropower projects were developed on a large scale; 175 GW of capacity with storage of 9,000 gigawatt hour (GWh) was created[7] (Annexe 3.C.).

The pump storage projects provide the ideal flexibility needed for maintaining grid stability. They can supply additional electricity promptly in the event of a sudden shortfall in expected generation, or an unforeseen spike in demand. Table 1 (Annexe 3.D.) gives a comparison of the flexibility that pumped storage provides compared to gas and coal-fired thermal power plants. The pump storage projects are clearly superior. Because of their long life and nominal maintenance expenditure, they are relatively cheaper in terms of life cycle costs though the initial capital cost is high.

Then off-river pump storage projects can be developed. These projects would be away from rivers; two reservoirs would be created with a height difference wherever the natural terrain makes this feasible. Water has to be brought from some source for the reservoirs. But thereafter no additional water is needed. Water can be pumped up from the lower reservoir when there is surplus electricity. This water from the upper reservoir can be released to come down and generate hydropower when needed. As these off-river pump storage projects do not involve the construction of dams and structures on rivers, they can be constructed in a shorter period of time. They can also be developed at mining sites after the mine has been exhausted. Some examples of off-river projects in abandoned mine shafts are given in Annexe 3.E.

Having put up a few pump storage projects earlier, India is now moving swiftly to develop pump storage projects. The pump storage projects potential on rivers in India was estimated to be 103 GW.[8] This number is likely to be revised upwards to over 180 GW. The identified sites would be adequate for meeting all storage needs in the near term.

The off-river potential is huge in India except in the flat Indo-Gangetic plains. Identification of potential sites using satellite imagery by a government agency would be the best way to assess this potential. This data can be given to potential developers for a token fee with a more substantive payment later when they begin developing a project at a site. This process can be self-financing. Open-cast coal mining sites, after the mine is exhausted and ready for closure, are also potential sites for the development of pump storage projects.

Private developers are seeing pump storage projects as an attractive investment opportunity. Investor friendly guidelines were issued by the Power Ministry in 2023 which simplified and eased the clearance process. The pace of development of pump storage projects has been gathering momentum. Developers have been quick to identify sites and initiate development of pump storage projects. Projects with a total capacity of over 65,000 megawatts (MW) have been taken up for project preparation/development. Pump storage capacity of 8,000 MW is under construction (see Annexe 3.F.). There is, as a result, growing confidence that India would be able to create storage on the scale required in the coming years.

Concentrated Solar Thermal Projects with Storage

In concentrated solar thermal projects with storage, [9] large mirrors are used to reflect and concentrate the sun's rays to a point to get intense heat. This heat can be used to produce steam which can drive a turbine to generate electricity as in conventional thermal power plants. This heat can be stored in molten salt and can be retained for many hours. This stored heat can be used to generate electricity when needed.

Not many concentrated solar power (CSP) plants were developed earlier as costs were relatively high. This technology has now become mature and cost effective. The plants that have been developed in the last few years have costs that make such plants an attractive large-scale storage option. In addition, they have a substantial additional cost advantage if developed next to a large conventional solar power project using solar panels as the transmission capacity would be fully utilized, carrying solar power in the day and stored solar thermal power at night. Sunlight for solar panels is equally suitable for concentrated solar thermal projects.

The costs of concentrated solar power projects have come down with improved designs. Earlier the large mirrors were mounted on high towers, and the sun's rays were focused to a point (Figure 1). The costs have been brought down by placing the mirrors on the ground and focusing the reflected rays of the sun to a point on a high tower (Figure 2). Instead of mounting mirrors on many towers, only one tower for receiving the reflected sunlight is needed. With only one tower instead of many to mount the mirrors, the costs are lower. The heat-retention capacity of molten salt has been improving with better chemicals in it. The US Department of Energy (DOE) was targeting a cost of

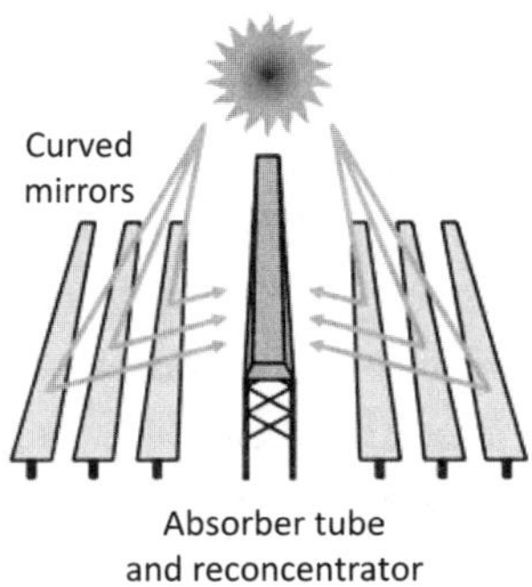

Figure 1 Linear Fresnel reflectors

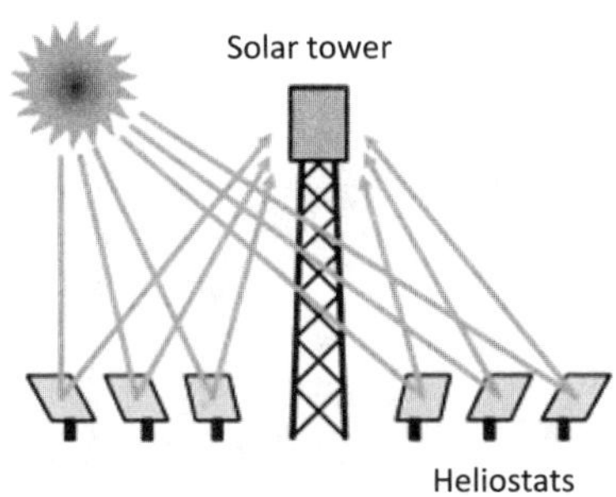

Figure 2 Central receiver system

5 cents per unit (kWh) with the R&D projects that it was supporting. With scale, the production cost of the mirrors and of molten salt would decline.

The 950 MW CSP-PV (concentrated solar power – photovoltaic) hybrid plant recently set up in Dubai provides solar power at 7.30 cents per kilowatt hour (kWh), a price competitive with fossil fuel-based power generation, on round-the-clock basis. This is a major inflection point for the world in the transition away from fossil fuels in electricity systems. A few countries have developed a significant number of concentrated solar power projects. (Annexe 3.G.).

When the Solar Mission was started in 2010, the potential of concentrated solar power was recognized. Bids for concentrated solar power without storage were invited. These were separate from the bids for solar PV projects. The bid parameter was the same, price per unit of electricity supplied. After the first round no more bids for concentrated solar power were invited as the price of electricity from solar panels began to fall rapidly. Some of the successful bidders who were selected for concentrated solar power could not develop their projects. The technology then was at an early

Figure 3 Noor Energy, Dubai

stage of development and the technology partnerships of the successful Indian bidders didn't work out well. But two of them did complete their 50 MW projects and are more than happy. Their operation and maintenance costs have been lower than they expected. The plants have been running very smoothly, giving them good returns.

India needs to begin developing concentrated solar power with storage at the earliest to supplement its PSP programme. The need for storage would keep rising as renewable energy capacity increases. Success in storage is the only way of moving towards a carbon-free electricity system. The decline in the cost of concentrated solar power with storage makes it a very attractive option. In the existing solar parks, land can be earmarked and bids invited for concentrated solar power. Solar radiation data may be provided to the bidders to enable them to make realistic cost estimates and give competitive price bids. The transmission capacities created for evacuating electricity from these solar parks would then be used round the clock and the real cost for the consumer would come down. Wherever solar panel projects can come up, concentrated solar power can also be put up. Concentrated solar power projects, however, need tailor-made designs after proper mapping of solar radiation to optimize the plant design and minimize the capital cost.

The capital cost of concentrated solar power projects is primarily in the mirrors and the structures on which these would be mounted. Both can be made in India. With a large programme, prices can be expected to decline considerably as happens usually with new products. It would therefore make sense for India to announce its intention to have a large programme as it did when it launched the Solar Mission. This enabled investors to prepare for the market that would emerge. This combined with repeated competitive bidding facilitated movement down the cost curve. This should be done for concentrated solar power projects now.

Battery Energy Storage Systems

Batteries have been in use for a long time. As mobile phones became popular and the size of the market grew exponentially, the development of battery technology saw a major breakthrough. The lightweight compact lithium–ion batteries were developed. These are used in all smartphones. Then electric vehicles (EVs) entered the market. The lithium–ion batteries became bigger and better with the capacity to enable cars to drive a few hundred kilometres in one charge. With an increase in battery capacity, heavier and more powerful vehicles, big luxury cars, buses and light-duty trucks, have electric versions on the market. The driving distance in one charge has been going up in newer models.

Batteries with larger capacities for grid storage became the next frontier. Though smaller batteries for mini grids using solar power for round-the-clock power supply have been in the market along with solar panels, grid-scale storage needed a quantum increase in capacity. There has been a dramatic breakthrough in the reduction of the cost of batteries and competitive supply for grid-scale deployment. There has been a 90% decline from the price level in 2010.[10] In recent years, the doubling of market demand has led to a 20% decline in prices. As many countries move towards making their electricity systems free of carbon emissions, they are moving rapidly to create storage. Battery energy storage systems (BESSs) are seeing a surge in demand.

In the last decade, laboratories and firms have been working on alternative battery chemistry and materials to increase capacities, lower costs, and dependence on scarce minerals. This has been driven by the legitimate concern whether the planet has the lithium reserves, critical minerals, and rare earth deposits to get all auto vehicles to become electric and meet the requirements of grid-scale storage for the world's entire electricity system to become net zero. Sodium, the next element in the periodic table after lithium, is available in abundance. Sodium–ion batteries have been developed. Cars using sodium–ion batteries are being sold in China for over a year. There is a startup from the Massachusetts Institute of Technology (MIT), called Ambri,[11] which with a radically different approach of using abundantly available materials has developed a 'molten metal' battery. Other routes show promise, and breakthroughs are likely.

India has been forward looking, creating a competitive market for battery energy storage systems. It has done so by extending the practice of competitive bidding with the price of the supply of stored electricity being the bid parameter in bids for round-the-clock supply. Battery energy storage systems became the choice for storage of many successful bidders who tied up battery energy storage system supply in the global market, which is highly competitive, and where costs have been declining rapidly. A Viability Gap Funding (VGF) scheme for accelerating the installation of battery energy storage systems was introduced in 2023 (Annexe 3.A.). The government had begun to see the need for speedy installation of storage to ensure optimal utilization of generation capacity during the hours of high solar power generation. The need for storage would rise in tandem with the increase in solar power capacity. Battery energy storage systems have the advantage of having a short lead time for installation; around 18 months from the date of the award of a contract. Battery energy storage systems are ideal for ramping up storage in the short term. There are no supply-side constraints, and these can be installed wherever needed. The Solar Energy Corporation of India (SECI) began the process of inviting bids for battery energy storage system. States have followed. Bids have seen a substantial downward trend in prices which has been continuing (Annexe 3.B.). Solar power with battery energy storage system is now cheaper than new thermal power.

India could well become a global leader in the deployment of large-scale grid storage.

The age of coal in India is over.

Consumer Choice for Real-time Zero-carbon Electricity Supply

To accelerate the reduction of emissions from electricity, DISCOMs across the country may be mandated to give consumers with a load of, say over 10 kW, the choice of buying green carbon-free electricity on a real-time and not net basis, from a prospective date. A prospective date is being suggested so that the DISCOM gets time to make arrangements for supply of zero-emission electricity; from renewable energy when it is generated and from stored renewable energy when renewable energy generation does not take place. There would then be a bottom-up demand push for the creation of storage. The cost for such supply would emerge as the DISCOMs arrange the procurement of zero-carbon emission electricity to meet demand. Based on this cost, there would be the need for a separate tariff for the supply of this carbon-free electricity. As demand growth and corresponding supply arrangements would be very dynamic, it would be better if the DISCOM is allowed to fix the tariff for real-time green electricity on its own, issue bills, and collect revenue. The State Electricity Regulatory Commissions (SERCs) may undertake subsequent scrutiny to satisfy themselves that procurement has been done efficiently, and that the consumer has not been overcharged.

It is likely that a large number of industrial and commercial consumers may opt for this to lower their carbon footprint, add to their brand value, and be part of supply chains wanting to sell in the EU without having to pay higher import duties under the Carbon Border Adjustment Mechanism (CBAM). A large number of individual consumers wanting to contribute to the battle against global warming may also choose to buy green electricity and be prepared to pay a higher cost. There is, fortunately, in India a momentum among industrial firms and civil society for becoming carbon neutral through their own choices and action. The DISCOMs need to be mandated to meet this demand from their consumers. This demand could grow quite rapidly.

Competitive Procurement

India has benefitted enormously by competitive procurement of electricity through long-term power purchase agreements which de-risk investment. As a result, contracted prices have been lower than expected. The law permits tariff determination: (a) on the basis of prudent costs after scrutiny by the Regulatory Commission, and (b) acceptance by the Regulatory Commission of the market discovered price through a transparent bidding process.

Policy has shifted decisively in favour of procurement through bidding.

The same approach would be the best for storage. This is being adopted for round-the-clock bids where the bidder chooses his storage option and gives a price bid for the supply of electricity; renewable energy with storage. This is also being done for stand-alone battery energy storage system projects.

For concentrated solar power with storage, inviting a price bid would not present a problem as it has already been done for concentrated solar power without storage in the past.

Hydro projects have costs which are site specific. Their tariffs continue to be determined by the Regulatory Commissions in accordance with their regulations. These give the prescribed reasonable return on investment. The costs of pump storage projects are similar to hydropower projects—site specific.

Developers are identifying sites and getting from the state governments the authorization to develop these sites. Each site has its own storage potential. Is competitive bidding for supply from different sites with varying capacities at all feasible?

One unorthodox way of seeking bids and getting the benefit of lower costs from competition would be to make the bid parameter the cost per unit of stored electricity and be agnostic about the size of the project. Bids would then be received for different capacities, and the bid for the lowest price per unit would be selected. With repeated bidding, low-cost sites would be developed first. Developers who have sites would see how they could lower their costs, or margins, or both to get contracts. This approach could be used for both river and off-river sites for pump storage projects. As the bids would be repeated, the DISCOMs would be able to contract the quantities that they need after a few rounds of bidding. They would get the benefit of lower prices from competition.

Reducing the cost of storage further would be likely if DISCOMs were to invite bids for storage capacity and not the supply of stored carbon-free electricity. The developer of storage has then a much simpler task of developing storage capacity, maintaining and ensuring availability for use when needed. This would in some ways be like the provision of transmission capacity. Without having to take on the responsibility of arranging green power for storage and supply for the long duration of his contract, the bidder's responsibility would be reduced and the bid price for storage would be lower. The DISCOM would also gain as it would be able to use its contracted power supply for storage in a flexible manner to minimize costs. It could even buy power from the power exchange for storage to reduce its costs. DISCOMs contracting storage capacities through competitive bidding should lead to significant lowering of storage costs. The commercial case for the energy transition would become stronger, the clearer it becomes that renewables-plus storage can provide reliable and cheaper power supply to meet additional demand.

The added advantage of DISCOMs contracting for storage capacity rather than supply of stored renewable energy would be that DISCOMs could get their contracted thermal plants to run at full load and use surplus power for storage to meet peak demand. From purely commercial considerations it may be best for DISCOMs to contract separately for storage capacities and for the supply of solar and wind power separately. Contracts for the supply of renewable power, and for storage separately can be calibrated by Discoms to ensure reliable supply for projected demand. The salience of round-the-clock supply contracts for carbon-free electricity may decline as a result.

Long-term Storage

The storage options of pump storage project, concentrated solar power, and battery storage provide daily storage. These can take care of demand at night as well as during the peaks in daily demand. But there is a seasonal dimension which needs to be recognized. This needs a distinct solution. Demand varies considerably over the year. In the summer when temperatures in the day in North India remain over 40 degrees Celsius, electricity demand for running air conditioners rises sharply. This demand is rising rapidly and will continue to do so. How to cater to seasonal peaks in demand? At present this is conceptually unchartered territory. Some combinations of the following options would have to be adopted.

Costs would have to be discovered.

a. Green hydrogen is produced without carbon emissions and is stored energy. It can be used as a substitute for fossil fuels. It provides long-term storage. It can be converted into green ammonia which can also be used as an energy source. Green hydrogen or ammonia can be used to generate electricity from a thermal turbine.

The power plant would have to be specifically designed for using green hydrogen or green ammonia as the fuel. These can be used to generate zero-emission electricity to meet seasonal peaks in electricity demand. Small pilot projects, one using green hydrogen and the other green ammonia, would be needed to find their costs to determine which one would be better and the preferred long-term solution.

b. India produces huge quantities of crop residues. The rice crop residue in North India is burnt in the fields and leads to a significant spike in air pollution. This can be converted into pellets. These can be used to generate thermal power in plants designed for them. The energy in them is renewable. There is no net increase in carbon emissions in using them for electricity generation. Hence, using these for generating electricity to meet peak seasonal demand is another option.

c. India would be creating substantial renewable energy capacities along with storage for producing green hydrogen as planned under the Hydrogen Mission. During the seasonal demand peak periods, hydrogen production could be reduced. This would enable the diversion of carbon-free electricity from renewable energy to meet the seasonal demand peaks.

Undertaking pilot projects soon for each of these solutions would be desirable. Individual technologies would be demonstrated; costs would be discovered; which option is cheaper would be known. Also the long-term potential of the options could be assessed. Capacities for scaling up would be created. Some ideas of the potential for cost reduction through scale and competition would also emerge. It would then be easier to choose the least-cost pathway for reliable supply of zero-emission electricity round-the-clock throughout the year. Estimation of seasonal peak demands, growth, and planning for meeting this demand would become increasingly important and critical for maintaining reliable supply.

Annexes

Annexe 3.A.: Viability Gap Funding Scheme for Battery Energy Storage Systems

To promote large-scale deployment of battery energy storage system, a Viability Gap Funding (VGF) scheme was started in September 2023 for three years. The scheme provides financial support for projects in a phased manner:

1. 10% of VGF is released at financial closure
2. 45% is disbursed upon achieving Commercial Operation Date (COD)
3. 15% is provided each year over the next 3 years from COD

According to the National Electricity Plan 2023, India will require 236 GWh of battery energy storage system capacity by 2031–32.

Annexe 3.B.: Declining Trend in Stand-alone Battery Energy Storage Systems' Prices

Tender month	Tendering authority	Power (in MW)	Energy (in MWh)	Winning quote (in Rs/MW/ month)*
22-Jun.	GUVNL	25	50	10.66 lakh
23-Nov.	GUVNL	250	1,000	4.49 lakh
24-Aug.	MSEDCL	500	1,500	2.19 lakh
25-Feb.	NHPC (AP)	500	1,000	2.08 lakh

Total awarded BESS capacity – 13.25 GW
*Contracts awarded post-September 2023 benefit from the VGF scheme.

Annexe 3.C.: Pumped Storage Capacity in the World

Around 175 GW of pumped hydro storage capacity was installed worldwide as of 2022, with leading countries being China with 44.7 GW, followed by Japan with 27.5 GW, and United States with 22 GW. Figure 4 presents the status of pump storage projects installed across countries having large storage capacities. [12]

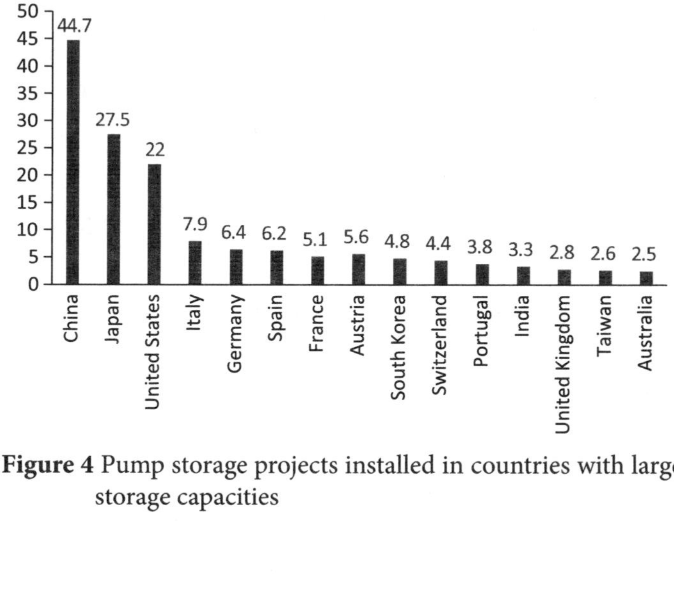

Figure 4 Pump storage projects installed in countries with large storage capacities

Annexe 3.D.: Advantages of Pumped Storage Plants

Pump storage projects have a lifetime of over 40 years, and they operate with an efficiency of 70–80%.

Compared to a thermal power plant, pump storage projects have better ability of quick start-stop as well as higher ramping. [13]

Flexibility capabilities of typical thermal and pumped storage plants

Plant type	Pumped storage	OCGT	CCGT	Coal-fired plants
Start-up time (cold start)	75–120 sec.	5–10 min.	120–240 min.	300–600 min.
Minimum load (% of P)	35–45%*	40–50%	40–50%	40-55%*
Average ramp rate (% of P/min.)	80–100%*	8–12%	2–4%	1–4%

*In Indian thermal plants
OCGT – open-cycle gas turbine; CCGT – combined-cycle gas turbine;
P – power output

Note: These features enable PSPs to provide multiple services to the power grid. The high-ramping capability helps it deal with the sudden increase of load in the power system. Furthermore, it can smoothen the sudden fluctuations in renewable energy generation and can also provide frequency and voltage support ancillary services.

Annexe 3.E.: Some Pumped Storage Plants in Abandoned Mines

In the US in New Jersey, a partially underground pumped storage hydropower project with a capacity of 1,000 MW was developed within a decommissioned iron ore mine 760 metres below the surface.

In California, the Eagle Mountain project utilizes two disused mining pits—serving as upper and lower reservoirs—to generate 1,300 MW of electricity.

In Spain's Asturias region, water from a flooded coal mine is used as a source for a semi-underground project.

South Africa has used an abandoned gold mine to develop a cascading project.

The development of the 200 MW Lewis Ridge Closed Loop pumped hydropower storage project at a site of a former coal mine in Kentucky has been initiated. [14]

China initiated project at the Shenhua Daliuta coal mine in 2010, where 32 underground reservoirs were constructed.

In Germany the abandoned mine shaft of the Prosper-Haniel coal mine was used to put a pump storage project. Germany plans to build a pump storage plant in Upper Harz at an abandoned mine with a capacity of 100 MW. [15]

Annexe 3.F.: Pump Storage Projects Development in India [16]

Status		Capacity (in MW)
Pumped storage projects' potential		1,83,000
Projects in operation		
Nagarjuna Sagar, Srisailam, Bhira, Ghatgar, Purulia[$]		4,700
Projects under construction		
Pinnapuram 1200, Tehri 1000, Kundah 500, U Sileru 1350, Sharavathi 2000, MP30 1920		8,000
Projects allotted by states for development		
	Projects concurred by CEA and yet to be taken up for construction Turga 1000, U Indravati 600, Bhivpuri 1000, Bhavali 1500	4,000
	Projects under survey and investigation	62,500
Total		**74,500**

[$] Details available at <https://cea.nic.in/wp-content/uploads/hpi/2023/05/Status_of_Pumped_storage_development_in_country-2.pdf>

Annexe 3.G.: Countries with Large Concentrated Solar Power Plants

Spain

Spain has about 2.3 GW of concentrated solar power plants operating as on 2023.

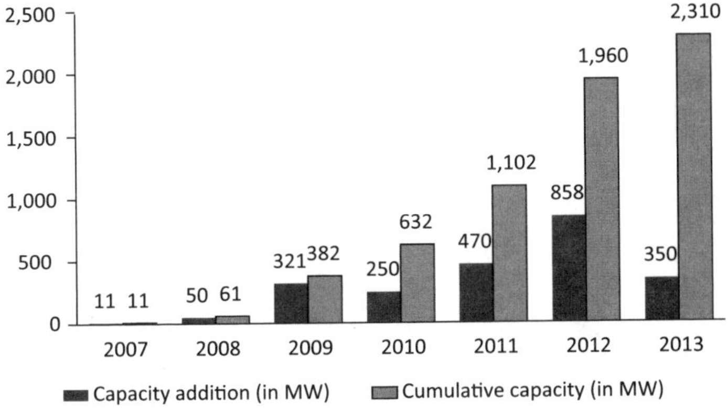

Figure 5 Growth of concentrated solar power capacity in Spain

USA

USA has about 1.5 GW of concentrated solar power plants operating as on 2023.

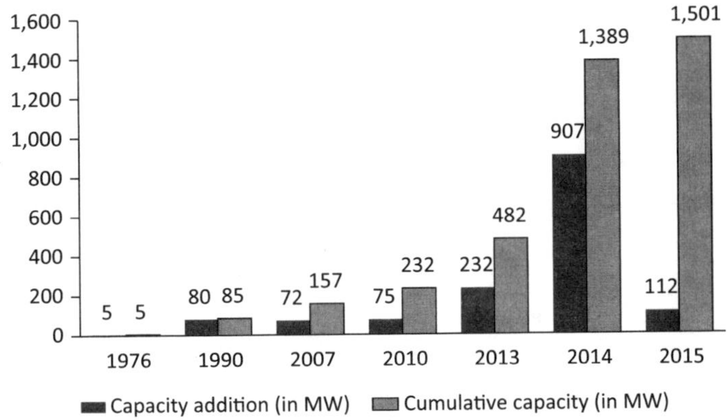

Figure 6 Growth of concentrated solar power capacity in the USA

China

China has about 596 MW of concentrated solar power plants operating as on 2023. In addition to this, 310 MW of concentrated solar power plants were under construction. Many more new projects are being developed.

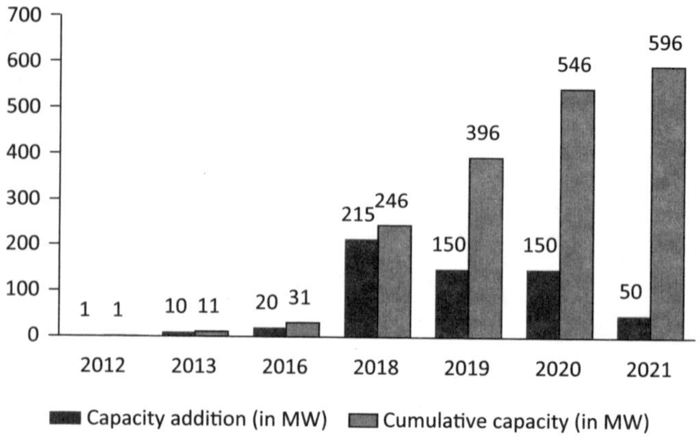

Figure 7 Growth of concentrated solar power capacity of China from 2012 to 2021

CHAPTER 4
Forging a Fossil-free Electricity System

Introduction

Creating 500 gigawatt (GW) of fossil fuel-free capacity, as announced by India at Glasgow during COP26[1] is not imposing additional costs. It is now the cheaper way for meeting the increasing demand for electricity.

India has reached a key inflection point in its energy transition. The cost of renewables in India is among the lowest in the world.[2] The cost of battery energy storage systems (BESSs) has been coming down. The life cycle cost of pump storage plants (PSPs) is low. The cost of carbon-free electricity from the renewable energy sources of solar and wind combined with storage, battery energy storage systems as well as pump storage plants, has become lower than that of new thermal power. Meeting all additional electricity demand from renewables along with storage is now the cheaper option.[3] The development of new thermal power plants is no longer necessary from purely commercial considerations (Annexe 4.A.).

This is a transformative development of the last few years. Increase in electricity generation and consumption is getting decoupled from increase in carbon emissions. This has created an altogether new paradigm. This needs to be fully understood, assimilated, and acted upon.

Misplaced Concerns

Some of the traditional views have got dated. Apprehensions flowing from these are misplaced. These are:

- India's right to development is paramount. Increasing affordable energy production and consumption is essential for this. India has abundant coal reserves, and this gives cheap electricity for economic and industrial growth. As India's per capita emissions are so low, India has the right to use more coal. [4] This right flows from the principle of 'common but differentiated responsibility' enshrined in the United Nations Convention on Climate Change (UNFCC).[5] But having the right does not mean that this right must be exercised even if it means choosing a more expensive option. Indian consumers have to unnecessarily pay more, the longer this right to build new coal-fired thermal plants is exercised.
- There are hidden costs of renewables which need to be factored in. If this is done, renewables are not as cheap as they appear. This is correct. A distribution company (DISCOM) has the added cost of supplying electricity to meet demand when there is no renewable energy generation, or it is inadequate. But by creating sufficient storage this additional cost would still make carbon emission-free electricity cheaper than new thermal power. This has been analysed and explained in the last chapter on 'storage'.
- Transmission costs are presently not being recovered from renewables. This is being met through a cross-subsidy regime in transmission. This was a good promotional measure in the early days when solar power was very expensive and quantities were minimal. This

waiver of transmission charges with the cross-subsidy regime is no longer needed and should be done away with. It is better for true costs to be reflected in prices. These would lead to better decisions on location of new investments. Renewables combined with storage paying normal transmission charges would still be cheaper than new thermal plants. The development of small decentralized solar power installations near load centres would gain momentum as true transmission costs get reflected in the price.

- Storage cannot be created at the scale needed to meet all additional demand. This apprehension is imaginary. The way to proceed would be to rapidly increase the quantum of bids for storage. So far response to bids has been good and prices have not only been competitive but are also coming down. There are no constraints for development of battery storage, off-river pump storage, and concentrated solar thermal projects. All of these take less time to put up than a coal-fired thermal power plant.
- The pump storage projects on rivers may have a larger gestation period. However, the pump storage projects' potential in the country is very large and now is estimated at 181.4 GW.[6] Fortunately, the response of private developers to this emerging market opportunity is very positive. Project preparation work for 74.5 GW has already been started.[7] With positive and investor friendly policies, these projects can be completed in a reasonable time. Technically, they are simpler to build than normal greenfield hydropower projects. Private sector project execution capacities have also been getting better and are now world class.

- India does not have the land for meeting all its future additional energy needs from renewables. It turns out that land is not a constraint. The solar power potential of the country is far larger than earlier assumed. According to one recent study the potential is 6,653 GW. By adding the emerging potential of agri-solar, the potential could go above 10,800 GW (Annexe 4. B.). In agri-solar, solar panels are installed on land being used for farming. There is a net increase in the farmers' income as they not only get income from their farm produce but also from the sale of solar power to the DISCOM. A few pilot projects are already under implementation in India. The potential is breath-taking.
- Sudden changes in the weather can lead to an unexpected fall in renewable energy generation. This is a legitimate concern. In dealing with the uncertainties of the weather and its impact on renewable energy generation, India is better placed. Hydropower capacity of 47,930 MW[8] constitutes a significant share of total generating capacity. Hydropower is intrinsically flexible. Then there is an integrated robust all-India electricity grid. Unforeseen and extreme weather events in some part of India and its impact would be mitigated by normal renewable energy generation in other parts. Having sufficient spare generation capacity and storage would be essential as the share of renewables rise.

Even if renewable energy with storage is cheaper, can India now meet all its growing demand for electricity from only renewables and storage? Can renewable energy and storage capacities be created on the scale and at the speed that would be needed? Yes, it can.

Private developers have the capacity for developing more projects and would respond with competitive price bids for solar, wind, and storage projects. Land is not a constraint for scaling up solar power project development. There are a number of solar parks where land is readily available (Annexe 4.C.). The identified pump storage potential alone is large enough for meeting storage needs in the near term. Battery storage systems can be supplied without any constraints. Hence all that is required is to increase the pace of inviting bids. The private sector would respond with competitive bids and undertake timely completion of projects as it has been doing till now.

Finance has not been a constraint so far and it is unlikely to become one going forward. Private developers only need tenders to bid for. With power purchase agreements (PPAs) with DISCOMs they can finance their projects, raising both equity and debt. The power purchase agreements with DISCOMs has enabled India to succeed in getting private investment in power generation. Acting on any misplaced faith in a theoretical paradigm of a more efficient market and doing away with this long-term power purchase agreements pillar of derisking private investment would be a grievous mistake. Private investment on the scale needed would not take place and power shortages would return.

The DISCOMs need to do their demand projections and then contract for renewables and storage to be able to provide reliable supply to meet all the increase in demand. Contracted supply should be adequate even when the economy and demand grow faster than expected. In undertaking demand growth projections, they also need to factor in the rapid pace at which electric vehicles (EVs) are gaining market share. A comfortable cushion of reserves and spare capacity should become the norm. This assessment should be central to the resource adequacy planning (RAP) exercise of the DISCOMs. The state regulators should subject this to transparent scrutiny and err on the side of creating extra capacity.

India is reaching a level of development where consumers would gladly pay more for greater reliability of supply. The weaker sections are being taken care of by the free supply of electricity by some states and subsidized supply by others. Now the central rooftop solar panel programme would provide 300 units of free solar power to households of the weaker sections. The rest can and would gladly pay what is needed for reliable quality power supply.

Solar Power

As elaborated in Chapter 2, India's success in the development of large grid-connected solar power projects has been remarkable, surprising observers in India and the world. This gave India the confidence to launch the International Solar Alliance (ISA) in 2015. This now has 120 countries as members and aims to give greater momentum to the use of solar energy, especially in the developing countries (Annexe 4.D.). Rapid increase in the use of solar power as a substitute for fossil fuels is urgently needed for the world to move to net zero.

India aims to have 500 GW of fossil fuel-free capacity by 2030. Solar power would have the major share of the additional capacity that would have to be created. Solar power capacity has already exceeded 100 GW. India has been developing solar parks where private developers could get land to develop their projects on successfully securing a contract for the supply of solar power in response to the bids that are being invited. More bids are being issued. The creation of 500 GW of fossil fuel-free capacity should not be difficult.

Rooftop Solar Power

India's solar power programme has had a distortion in favour of large projects. About 89 GW has been installed through large grid-connected solar projects.[9] However, it has only about 22 GW, from decentralized small solar panel installations on rooftops and in rural areas though these have the inherent advantage of not needing transmission investments.[10] This imbalance is now seeing rectification. Captive solar panel installation by consumers is rising.

The rooftop solar programme for provision of 3 kilowatt (kW) solar panels to 1 million households started in 2024 by Prime Minister Modi is a landmark initiative.[11] These are expected to provide around 300 units of electricity in a month. In this programme, two-thirds of the cost of the panel is a grant and the balance is on credit. It is probably the most comprehensive provision of lifeline green carbon-free electricity to low-income households in the world.

Under this programme, 30 GW capacity creation would take place. Competitive democratic politics is likely to lead to the extension of this programme to cover all similarly placed households. If 50 million households are covered over, say, the next 10 years, capacity creation of 150 GW would take place.

Solar Power for Pump Irrigation

Irrigation by using electricity to pump groundwater has enabled India to keep increasing food production and remain self-sufficient for a growing population which, at over 1.4 billion, is now the largest in the world. One of Prime Minister Modi's major initiatives has been the programme for substituting power from solar panels for grid-supplied electricity for irrigation. This programme called Pradhan Mantri Kisan Urja Suraksha evam Utthaan Mahabhiyan (PM-KUSUM) is transformative (Annexe 4.E.). Around 20% of India's total electricity consumption is for irrigation. [12] With this programme, it would move away from fossil fuels to solar power. Solar panels and energy-efficient pumps are being given to farmers. Two-thirds of the cost is a subsidy being shared equally by the central and state governments, with loans being provided for the balance one-third of the cost. Solar power that is not needed for irrigation by the farmers is to be purchased by the DISCOM. The income of farmers would increase. As the programme gets completed and all irrigation is done by solar power, about 150 GW of solar power capacity would have been added.

The programme would gain greater momentum if the design were to be changed and a state agency or the DISCOM is given responsibility, rather than leaving it to the farmers to take decisions and get the solar pumps and seek the subsidy benefits, as is the case at present. With bulk procurement of both solar panels and energy-efficient pumps the costs would also become lower. Subsidy and debt may be tied up and the farmer may be approached to say yes and have the solar panel and pump installed by the state agency without making any effort. A proactive role by DISCOMs enabled India to achieve a remarkable breakthrough in the use of energy-efficient LEDs in place of incandescent bulbs. This has been a major pillar of India's achievements in energy efficiency. The replication of this approach for the implementation of PM-KUSUM would see a rapid acceleration in the replacement of electricity from the grid by the farmers' own solar power for irrigation.

This programme would also result in a substantial improvement in state finances. At present, the subsidy being given by the state governments every year for the supply of free or/highly subsidized power for irrigation to farmers is around Rs 80,000 crore. With the switchover to solar power for irrigation, this subsidy burden on the state governments would decline and end when all irrigation is done by solar power.

If farmers get paid for the solar power that they do not use for irrigation they would have an incentive to be efficient in the use of electricity. The consequential benefit would be that they would use water more efficiently. Water-use efficiency is urgently needed. India is mining groundwater whose levels are falling. If course correction does not take place, the consequences would be catastrophic. But finding a policy instrument which would work has been difficult as cost-reflective pricing of electricity for irrigation has not been politically feasible. This would also open up the possibility of farmers switching to crops which use less water. If the net income from an alternative crop using less water and hence sale of more solar power increases, a farmer would shift to the other crop.

Decentralized Solar Power

The cost of supplying a unit of electricity in the rural areas for the DISCOMs is the highest as supply has to be taken through long distances to small, scattered loads in village settlements and farms spread over large areas. If a DISCOM decided to offer to buy solar power in the kW range from villages at a price which gave a good return on the investment in putting up the solar panel, say, Rs 4.50 per unit, this would result in the DISCOM saving on a very conservative estimate over Rs 3 for every unit of electricity it provides as its cost of supply is well over Rs 7.50. The DISCOM could buy from the consumer supply point, or at the distribution transformer to the extent it can technically absorb solar power. There is, therefore, a strong case for the DISCOMs to do so. The announcement of a feed in tariff by DISCOMs with the approval of the State Electricity Regulatory Commissions (SERCs) in each state is the way forward.

India has around 6,00,000 villages. Assuming a modest 1 MW solar power installation in rural areas with panels over the homes of farmers and over the areas where their cattle, fodder, fertilizer and foodgrains are kept, there is a 600-GW of solar power capacity that would come up over time. This would not need any subsidy from the government. A farmer could buy and install solar panels on his own. He could do this from his savings, or by taking a loan, or getting the solar panel on instalments; the money for paying the instalments coming from the sale of solar power. Alternatively, he could just lease out space for local businesses that may emerge, and which would install the panels and sell electricity to the DISCOM.

This would create dispersed jobs in rural India for installing solar panels. It would also increase the income of farmers. For the larger national goal of increasing the income of farmers this is a costless option with no risk. As the programme takes off, the DISCOMs would save increasing sums of money in buying solar power and using it to meet increasing demand in rural areas.

This dispensation should be applicable to the entire area of the DISCOM. In urban areas this should replace the existing provision of net metering. Net metering discriminates in favour of the better off who buys electricity at the highest tariff slab in the present cross-subsidy regime. They get paid for the supply of electricity from their rooftop solar panel to the DISCOM at the rate at which they buy. And this is higher than the lower rates which small consumers would get as they are in a lower tariff slab. Captive and rooftop solar power has become so cheap that their installation would take off. No subsidies are now needed (Annexe 4.F.).

Manufacturing of Solar Panels

India's solar power programme has succeeded using solar panels from China.[13] Chinese success in reducing the cost of the production of solar panels has been amazing. The price of solar panels from China went down dramatically and has made solar power commercially competitive. This is facilitating the transition away from fossil fuels in the world.

India would like to produce solar power panels as part of its quest for success in manufacturing. Its Production-Linked Incentive (PLI) programme under which grants are given to firms for achieving pre-agreed goals of investment and production is now supporting the production of the value chain of solar panels.[14]

If India really wants a breakthrough in manufacturing in a few years, it needs to adopt a different and bolder approach. It should use the policy of assured procurement by government to get private investment in India for full-value addition, from polysilicon to ingots, to wafers, to cells to modules. The way to proceed would be to first accept that solar panels made with full value addition in India may initially be more expensive than imported panels and that this would be acceptable. With increasing scale of domestic production, costs should come down.

The following would lead to success and self-reliance in less than a decade:

- Bids may be the invited by government for the supply of, say, 1.5 GW, the production capacity of an economical plant, of solar panels annually for 10 years with full-value addition in India from the year from which the new plant of the successful bidder would go into production. This assured long-term purchase commitment would completely de-risk investment. Competitive price bids are bound to be received provided potential bidders from India and overseas get enough time to plan to tie up the technology and the supply of capital goods for the manufacturing plant.
- To reduce the cost and risks for bidders, land in an Industrial Park may be offered at a reasonable rate along with environment clearance and its conditions as a part of the bid process. Cheap electricity supply from an older hydro or a thermal plant with the central government may be offered as a part of the bid. This has been done in the past for energy-intensive industries in the early days of India's industrialization. Energy constitutes a major part of the cost of production and prices would be lower to the extent supply of reliable cheap electricity is assured. Further, duty-free imports of capital goods as well as inputs for the production process may be offered as a part of the bid conditions.

- In addition to the lowest bidder, L1, if the next L2 and/ or L3 are willing to match the L1 price they may also be given contracts for the supply of solar panels of a smaller quantity, say, 1 GW each. The intention should be to create a competitive industry structure with at least three manufacturing firms.
- These bids may be repeated annually to get firms to compete and reduce prices.
- These panels may be used for government-subsidized programmes: for the solarization of irrigation under the PM-KUSUM programme and for the rooftop solar programme for 10 million households.
- Then these panels may be installed in all government institutions such as the railways, defence establishments, government offices and housing complexes, and campuses of public educational institutions.
- Special-purpose vehicle (SPV) may be created which would buy these panels from government. This special-purpose vehicle may in turn give the panels onwards to government agencies who would pay for these panels in instalments from the savings in their electricity bills. The number of instalments should be so fixed that the savings in the electricity bill would be enough for the payment of the instalments. The special-purpose vehicle would need working capital which government could arrange. The advantage of this approach would be that the government would not have to make budgetary provisions for these purchases. The normal fiscal constraints would not be a limiting factor for the programme.

With competitive procurement the prices of solar panels should come down to reasonable levels. These firms should start succeeding in the domestic market on their own. Self-reliance would become a reality. This would also enhance energy security.

Wind Power

Wind power potential is limited to the states of southern and western India. As technology advances and higher wind turbines can be put up, the estimated wind power potential rises. At a height of 125 metres, it is 695 GW and at a height of 150 metres, 1163 GW.[15] So far 51.6 GW has been developed.[16] The balance is getting developed with the invitation of fresh bids for wind power. The quantum for which bids are being invited needs to increase.

Repowering

There is also the significant potential of replacing the first generation of wind turbines that have been installed by bigger and more efficient state-of-the-art turbines with higher capacity to generate more energy at the same site. The potential through such replacement is estimated at 25 GW. [17] There has, however, not been much progress on this as the owners of the wind farms are getting money without any operating expenses and are not seeing the need to make fresh investment to pull down the existing wind power turbines and put up new expensive ones. The policies announced so far have not, yet had the desired response.

One unorthodox way of proceeding could be for these state governments to create SPVs. The SPV may buy out the existing wind farms and their contracts from the owners, with legislation, if necessary, with the commitment that the SPV would keep paying the present owners what they would be earning under the existing arrangements for the remaining period of their existing contracts. The owners should not be put to any financial loss. They would then willingly cooperate. The SPV may then invite tariff-based bids for the installation of wind turbines of the higher capacities that are now feasible at these sites. The existing turbines would be pulled down and new ones with higher capacities would be out up. The special-purpose vehicle would then sell the electricity from these wind farms to the state DISCOMs at a rate which would be cheaper than electricity from a new thermal plant. From this sale, it should be able to pay the original owners and get a small margin over and above its costs. The special-purpose vehicle would need only seed capital for buying out the existing wind turbines. But it should not need any subsidies from the state government. Re-development could be completed in a few years at no financial cost to the government.

Offshore Wind Power Development

In a forward-looking move, the government has initiated the development of offshore wind farms. The potential of offshore wind power is around 35 GW in Tamil Nadu, and 36 GW in Gujarat, according to initial assessments.[18] Feasibility studies have been undertaken. As it would be more expensive, guidelines for provision of viable gap funding (VGF) have been issued.

It may be advisable to take up one site with high potential, complete site investigation at public expense, plan transmission evacuation for the full capacity to be developed in phases. Thereafter, for the first-phase bids may be invited with site investigation data being made available to bidders. The site data should make it easier for the bidders to offer lower prices as they would not need to factor in the uncertainties of estimating their costs in the absence of precise data. As in the case of the Solar Mission, it would be desirable to create a competitive industry structure by awarding projects to a few developers at the same time. The responsibility for building the first phase of transmission may be given separately, preferably through competitive bidding. The cost of this power for the DISCOM after adding the transmission cost would naturally be higher than from onshore locations. The actual cost would be known after bidding. There would be learning and capacity building in executing these projects. The second round of bidding may be undertaken after the execution of the first round of projects is nearing completion. With the experience of project execution, cost reduction and lower prices could be expected in the second round of bids. The initial quantities of expensive offshore wind power would be miniscule. Their purchase could be made obligatory under the renewable purchase obligation if necessary. Depending on the prices emerging through competition a rational view could be taken thereafter; whether to rapidly scale up the pace of offshore wind power development or to do it more gradually.

Nuclear Power

India is now one of the cheapest producers of reliable nuclear power plants. It has, over the decades, developed the capacity to design and build nuclear power plants on its own. This development of capacity in a frontier technology area was a major challenge. India faced an international regime of technology denial and so had no option but to learn to design and manufacture the complete supply chain of all materials and components. Indian nuclear power plants are among the cheapest in the world. These plants after initial teething troubles now run smoothly.

Earlier, nuclear power generation in India also faced the constraint of not being able to import uranium to produce the fuel needed for generating nuclear power. Its own uranium reserves are modest. This changed with the Indo–US Nuclear Deal in 2005[19] and the subsequent waiver for imports from the Nuclear Suppliers Group (NSG) in 2008.[20] India is now building the next generation of 10 nuclear plants of larger 700 MW units for the first time.[21] Unit sizes had earlier gone up from 200 MW to 500 MW. But the existing capacity of 8.18 GW of nuclear power out of a total installed capacity of 475 GW is miniscule.[22]

Nuclear power generation can now be scaled up rapidly. Domestic manufacturing capacity would expand quickly to meet supply schedules as more orders are placed. In the coming decades nuclear power would have an increasing share in total electricity generation. There are no carbon emissions from nuclear plants. They provide stable electricity. Generation is not dependent on the uncertainties of the weather as is the case with renewable power. There is fortunately growing recognition in the world that nuclear power can replace fossil fuels in electricity generation and is also a solution to the challenge of global warming. Hopefully, it would start getting greater salience in the programmes for achieving net zero. There is some merit in the view that, as the share of reliable stable power supply from nuclear power begins to rise, closure of existing thermal plants can take place with greater confidence about the reliability and stability of power supply. Recognizing the potential of the role of nuclear power in its journey to net zero, India has recently announced its ambition of creating 100 GW of nuclear power capacity by 2047. To achieve such a large nuclear power capacity, it has decided to open up nuclear power generation to private investment by amending the Atomic Energy Act.

India is building its first fast breeder reactor which is expected to start functioning soon.[23] This would be a breakthrough in the quest for the reprocessing of nuclear fuel. When this happens India would have abundant fuel supply.

As nuclear power from domestically manufactured nuclear plants is price competitive vis a vis new thermal power, this ambitious nuclear power programme is a cheaper alternative to building new coal-fired power plants for generating electricity. These nuclear plants would be able to raise finances, both equity and debt, on the basis of their long-term power purchase agreements with DISCOMs. For DISCOMs, nuclear power would give them greater confidence that they could do away with fossil fuel-based thermal power. The programme would not need any financial support from the government as is now the case with all power projects.

India sees the emerging potential from small modular reactors (SMRs). The government has decided to finance the development of SMRs.[24] The Indian private sector is being associated with the effort. The intention is to design, develop, and manufacture SMRs which are globally competitive in performance as well as in price. Given the existing design and production capabilities for the supply chain of nuclear reactors, the development of an India SMR seems a realistic goal. On achieving success, these would facilitate the acceleration of nuclear power growth in India. Further, these could be exported to developing countries and become part of the global solution to achieving net zero by replacing fossil fuels in power generation. The developing countries do need to move towards net-zero speedily.

Taking Carbon Emissions to Zero

For India using imported gas, liquefied natural gas (LNG), for power generation has been too expensive in comparison to power generation from cheap domestic coal. India has been sensible in not looking at replacing coal with expensive imported gas for electricity generation as a way of reducing carbon emissions. Using expensive imported gas, as a fuel does not make sense now as the need is to move away from all fossil fuels at the earliest.

India has pursued successfully energy efficiency to reduce carbon intensity per unit of its gross domestic product (GDP). It has been increasing the share of renewables rapidly and reducing carbon emissions per unit of electricity.

Building new thermal plants is now not necessary. Renewables with storage is the cheaper way to meet the additional demand for electricity. When the commissioning of new thermal plants comes to an end, carbon emissions from electricity would peak. Thereafter, the per unit carbon emissions from electricity would decline as demand would rise and be met from non-fossil fuel resources. However, the total emissions would remain at a plateau till such time as the retirement of the existing stock of thermal plants would be taken up.

To get to zero emissions from electricity generation, the retirement of all thermal plants would have to take place. This would be a major policy decision and a critical milestone in the journey to net zero. There would need to be confidence about being able to ensure reliability of power supply. A decision on the appropriate, sustainable, and affordable pace of retirement would have to be taken. There would be a very large decline in revenues of both the state governments and the central government as thermal power generation is wound down. But there is no alternative to closing these plants to get to net zero. India would in the coming decades have to take hard decisions on closing its thermal power plants and of the pace at which it would do it. An understanding of these costs, informed discussion among stakeholders, and transparent consensus building well ahead of time would make it easier to evolve the future trajectory of the closure of thermal plants. This would have to be sustainable and just.

Cost of Winding Down Coal Production

India produces about 1 billion tonnes of coal and production is rising. It is expected to reach 1.5 billion tonnes by 2029–30.[25] The bulk of the coal produced is used in thermal power plants. Coal mining for thermal power plants would have to decline and then end as thermal power plants close. Coal mining for other industrial sectors would decline and may not be required at all as industries succeed in shifting away from the use of coal in their industrial processes.

The major implications of the end of coal mining would be:

- Jobs: Coal India employs 2,29,000 persons, excluding 1,11,124 contractual workers.[26] There are local communities whose livelihoods depend on the coal economy. These jobs would cease to exist. The local communities which are dependent on the coal economy would have to get alternatives which do not make them worse off, at least. This is the essence of the spirit of a 'just transition'.

- State finances: Royalty from coal mining is a major source of revenue for the central and state governments. This was around Rs 23,000 crore in 2022–23.[27] With the end of coal mining, these royalty revenues would end. These royalties constitute a substantial portion of the finances of the coal belt states of Jharkhand, Chhattisgarh, and Odisha.
- Central government finances: The central government levies a cess of Rs 400 per tonne for the use of coal. It is in effect a 'carbon tax'. The revenue from the coal cess in a year is around Rs 7,000 crore and would rise with increase in coal consumption.[27] This revenue would fall and become nil as coal consumption declines and ends. Coal India is a highly profitable company. It generates high profits, pays taxes on its profits, and dividend to government. This revenue stream would also end.

Reuse of Land

In preparing and planning for a sustainable and just transition, the coal industry is fortunate in having one very valuable asset, land. This can be intelligently monetized. This monetization could partly, or even substantially, cover the costs of a just transition. It could even generate a surplus. Unlike most countries India is short of land as it is densely populated. It has one-sixth of the world's population and only 2.4% of the world's land.[28] Land in India is not only expensive, but land prices keep rising. This makes land a far more valuable resource than it is in most countries as India is still at an early stage of industrialization and urbanization. As India moves ahead it will keep needing more and more developed land.

The coal companies have 2.94 lakh hectares of land.[29] If repurposed in a planned manner, this land can generate both money and jobs. New more productive and better paying jobs can and should be created by attracting private investment. To the extent, there is success in this, those working now in and around coal mining would become better off. Also, India has a long development journey ahead where per capita incomes would be rising. It should therefore be possible to create better livelihoods with higher incomes for those presently dependent on the coal economy.

The potential uses of land include:

- Development of new industrial townships: India is urbanizing rapidly as the pace of its development has gathered momentum. Non-availability of affordable land with infrastructure is a constraint. The need for new planned cities is real. However, developing new cities is not easy. After assembling land and developing physical infrastructure, there is the difficult challenge of getting investment which would create jobs. This in turn would create demand for housing and social infrastructure of schools, health care, and shopping complexes. There is, fortunately, the recent experience of developing new industrial areas as a part of the Delhi Mumbai Industrial Corridor project. Other corridors are also being developed. The key would be to get an anchor investor(s) who would generate employment. Once a few come in, then growth acquires a self-sustaining momentum. Coal mining areas are in relatively underdeveloped regions which need to accelerate the structural transformation of their economies.

- Tourism resorts: Coal mines exist in forests with scenic beauty. These can be developed imaginatively. With its growing middle class with rising incomes there is the need in the market for new attractive and affordable tourism destinations. Tourism is a labour-intensive service industry. The potential is huge. But to succeed this would need imaginative planning and marketing.
- Off-river pump storage projects: Twenty coal mine sites have already been identified for feasibility assessments and evaluation of pumped storage potential by Neyveli Lignite Corporation (NLC).[30] These projects leverage the unique topography created by coal mining, where two reservoirs at different elevations can be developed to store water and generate electricity through a closed-loop hydro system. Off-river pumped storage offers an attractive option to use mining sites where all the coal has been mined. Off-river pump storage projects need speedy development as large storage capacities have to be created to generate confidence that reliable supply can be assured without building new thermal plants

For successful implementation of these ideas, the state governments would have to take ownership. They have to provide leadership and regulatory support for land-use changes. Coal companies own the land. They are, at present, cash rich. They need to evolve creative strategies and partnerships for redevelopment and use of land for productive purposes. The central government could further enable this by establishing or mandating an existing development finance institution (DFI) to offer long-term capital. Such patient finance would yield long-term positive returns, not just financially, but also through the creation of modern, productive economic activity. As land is put to new productive uses, this would begin to generate employment, income, and tax revenues from production and consumption. This could gradually offset the fiscal losses from declining coal royalties and help the state governments manage the transition smoothly.

Winding Down of Thermal Power Generation

Thermal power companies have a choice. They can reengineer themselves as green energy companies, survive and prosper. Or, they can decline and fade away as thermal power generation comes down and then ends. The energy needs of India would rise manifold and therefore there is enough opportunity. Since the winding down of thermal power generation is many years down the road, they have ample time to prepare and undertake the transition to becoming successful green energy companies. Solar, wind, pump storage projects, concentrated solar projects, green hydrogen, and biowaste offer huge potential. But if they are unable to think and act strategically and begin transforming themselves, they would shrink and have to ultimately close down. They are profitable enterprises giving returns to their shareholders and taxes to the government. This would decline and then end unless they create new lines of business through diversification. Their management would be irresponsible if they decided to leave for their successors the challenge of developing the strategy and starting their transition away from fossil fuels. There is need to recognize that large organizations have greater inertia which makes it difficult for them to diversify. The government has to take a long-term strategic view and guide the transformation.

Thermal power plants near cities have highly valuable land. These lands can be redeveloped for urban uses. Such redevelopment is highly profitable. Some private firms have closed their old industrial units in Mumbai, developed new ones at new locations where land is relatively cheap and redeveloped their land in Mumbai and made handsome profits.

Plants at other locations can in theory be refurbished for generating electricity from concentrated solar projects with molten salt as storage plants. Being used for nuclear power generation is another option. Nuclear reactors, small modular reactors, and concentrated solar projects supply heat which is used to produce steam to run a conventional thermal turbine and generate electricity. The feasibility and viability of these conceptual options would need to be explored, and their feasibility demonstrated through pilot projects. If these efforts succeed, then the transformation of thermal plants to producers of zero-emission electricity could become a reality.

Retirement of Thermal Plants

For the retirement of thermal plants, the following sequence suggests itself:

- Old energy-inefficient plants in the airsheds of Northern India where air pollution is the cause of serious adverse health consequences should be closed first. The Badarpur coal-based thermal plant in Delhi was closed. This closure gave relief. A list of such plants can be drawn up, followed by taking up their retirement. Relief from air pollution should get overriding priority. The health benefits would justify immediate closure. Closure should have no adverse effect on power supply as so much new capacity is coming up with power flowing into the robust all-India electricity grid. The land of these plants can be reused, generating high profits.

- A list of old thermal plants which burn more coal per unit of electricity due to their obsolete technology was drawn up for closure by the Central Electricity Authority (CEA). The move was shelved on account of concerns about ensuring reliable supply. This closure can now be taken up. Regulatory changes can be made that mandate dispatch from the most energy-efficient plants first and the least-efficient last. This is easy. On the aggregate, less coal would be burnt for the same quantity of electricity generation. This would result in savings for the consumer. Carbon emissions per unit of thermal power being generated would also decline, as they do with all energy-efficiency gains.
- As the price of coal and railway freight rise, the variable cost of thermal power plants located far away from coal mines may become higher than the price of solar power. When this happens, it would make commercial sense for these plants to reduce generation when the sun shines. If the cost of solar power with storage becomes lower than the cost of coal and freight, it would make commercial sense to close these thermal power plants altogether.

When the big decision is taken to begin retiring all thermal plants, the same sequence of closing the oldest generation of plants first, would be logical. The longer the distance to which coal is carried to a thermal plant, the higher is the cost of generation. It would, therefore, make sense to consider closing these before taking up the plants of the same vintage located closer to the coal mines.

It would be for government to take a call on when to begin the retirement of thermal plants and the pace at which to do so. These would be difficult decisions. Stakeholder consultation and consensus building would be beneficial. There should be a lead time of a few years of preparation before closure of plants is started. The initial pace of retirement may be modest. With experience and learning the pace can be increased. The challenging part would be to ensure a just transition for the workers and the communities dependent on thermal power and coal mining. With advance planning and preparation, this should be ensured. Ample resources must be provided for this.

Technically, it should be possible to complete retirement in 10–15 years after the decision to begin retirement is taken. The transformation to a carbon-free electricity system seems to be more feasible than it would appear at first sight. An emission-free electricity system would be the major milestone in the journey to net zero. The earlier it is achieved the better it would be. It would generate greater confidence in the nation about achieving net zero.

To sum up, forging a fossil fuel-free electricity system is feasible.

It is affordable (Annexe 4.G.).

It may even turn out to be profitable if land redevelopment is done smartly.

Annexes

Annexe 4.A.: Cost Comparison of Solar-plus Storage with New Thermal Power

If solar power plus storage or thermal power from a new NTPC plant, Gadarwara Super Thermal Power Station, were to be used to meet the full demand of electricity in Delhi for a particular day, a simplified cost comparison would be as follows:

Gadarwara TPP tariff = Rs 5.33/kWh

Solar power = Rs 2.5/kWh

Storage cost (PSPs) = Rs 3.81/kWh

Storage cost (BESSs) = Rs 2.63/kWh

Power source	Tariff/blended cost (in Rs/kWh)
Solar + Storage (BESSs)	Rs 3.67
Solar + Storage (PSPs)	Rs 4.20
New coal-powered TPPs	Rs 5.33

Annexe 4.B.: Higher Solar Potential

A recent TERI report titled *Reassessment of Solar Potential in India: A Macro-level Study*[31] has concluded that on a conservative estimate the total solar potential of the country is 10,380 GW.

Category	**Solar potential (in GW)**
Ground-mounted solar PV on barren and unculturable land	4,909
Floating solar PV	100
Rural rooftop solar PV	600
Urban rooftop solar PV	360
Agri-PV for horticulture crops, coffee, and tea plantation	4,177
Rail and road solar PV	154
Building integrated PV	309
Urban PV	221
Total	**10,830**

Annexe 4.C.: Solar Parks[32]

A Solar Park is a large parcel of land developed with all necessary infrastructures and clearances for setting up solar projects. The Scheme for Development of Solar Parks and Ultra Mega Solar Power Projects was started in December 2014 with an aggregate capacity of 20,000 MW. Further, the capacity of the Solar Park Scheme was raised from 20,000 MW to 40,000 MW in March 2017 to set up at least 50 Solar Parks.

- The capacity of the Solar Parks is generally 500 MW or more.
- Approximately 4–5 acres of land is required for each MW of solar power.
- Rs 20 lakh per MW or 30% of the project cost, including grid-connectivity cost, whichever is lower, is provided as a grant to the state government by the Centre.
- The total Central Grants approved under the Scheme is Rs 8,100 crore.

Progress of Solar Park Scheme

- 55 Solar Parks with a cumulative capacity of 39,958 MW in 13 states have been approved.
- 18 Solar Parks with a capacity of 10,856 MW have been fully developed.
- 6 Solar Parks with a capacity of 4,775 MW are under development.
- Solar Projects with a total capacity of 12,209 MW have been commissioned in 24 parks.

Annexe 4.D.: International Solar Alliance[33]

- The International Solar Alliance (ISA) was launched by India and France in 2016.
- Its aim is to unite 'efforts to combat climate change by implementing solar energy solutions'.
- Over 100 countries have become signatories of the Alliance.
- The ISA has facilitated over 9.5 GW of solar applications in 55 developing countries, including Least Developed Countries (LDCs) and Small Island Developing States (SIDS).
- Nearly 4,000 people across 78 countries have been trained and certified to build local solar capacity.
- ISA has implemented demonstration projects to showcase scalable solar solutions.
- Some notable projects of the ISA include:
  - » Solarization of the Parliament building in Malawi
  - » Solar-powered cold storage in Seychelles
  - » Solarization of rural health care centres in Fiji and Comoros
  - » Solar-powered school in Kiribati
- The ISA has also established Solar Technology Application Resource Centres (STAR-C) in five countries: Ethiopia, Somalia, Kiribati, Cuba, and Côte d'Ivoire.
- ISA's Global Solar Facility aims to mobilize $100 million to support solar investments, particularly in Africa.
- By 2030, ISA aims to:
  - » Double installed solar capacity in LDCs and SIDS
  - » Provide electricity access to 18 million people

- Create 3.7 million jobs
- Mobilize a $10 billion multiplier effect to facilitate industry and economic development
- Unlock $1 trillion in solar investments by 2030

Annexe 4.E.: PM-KUSUM [34]

The Pradhan Mantri Kisan Urja Suraksha evam Utthan Mahabhiyan (PM-KUSUM) Scheme aims to provide solar power for pumping water for agriculture. It was started in March 2019.

The PM KUSUM Scheme has the following components:

- Component A: Setting up 10,000 MW of decentralized ground/stilt-mounted grid-connected solar or other renewable energy-based power plants by the farmers on their land.
- Component B: Installation of 14 lakh stand-alone solar agriculture pumps.
- Component C: Solarization of 35 lakh grid-connected agriculture pumps including feeder-level solarization.

Salient features

- Component A
  - » Renewable energy power plants (REPPs) of capacity 500 kW to 2 MW to be setup by individual farmers/ group of farmers/cooperatives/panchayats/ farmer producer organizations (FPOs)/water user associations (WUA) hereinafter called solar power generator (SPG). If they are not able to arrange equity required for setting up the REPP, they can opt for developing the REPP through developer(s) or even through local DISCOM, which will be considered as SPG in this case.
  - » DISCOMs will notify sub-station-wise surplus capacity which can be fed from such REPP to the grid and shall invite applications from interested beneficiaries for setting up the solar energy plants.

- » The solar power generated will be purchased by DISCOMs at a feed-in-tariff (FiT) determined by the respective State Electricity Regulatory Commission.
- » DISCOM would be eligible to get PBI @ Rs 0.40 per unit purchased or Rs 6.6 lakh per MW of capacity installed, whichever is less, for a period of five years from the commercial operation date (COD).

- Component B
    - » Individual farmers will be supported to install stand-alone solar agriculture pumps in off-grid areas, where grid supply is not available.
    - » Central Financial Assistance (CFA) of 30% of the benchmark cost or the tender cost, whichever is lower, of the stand-alone solar agriculture pump will be provided. The state government will give at least a subsidy of 30%; and the remaining at most 40% will be provided by the farmer. Bank finance can be availed by farmers, so that they have to initially pay only 10% of the cost and remaining up to 30% of the cost as loan.
    - » In North–Eastern states, Ladakh, Jammu and Kashmir, Himachal Pradesh, Uttarakhand, Lakshadweep, and Andaman and Nicobar Islands, CFA of 50% of the benchmark cost or the tender cost, whichever is lower, of the stand-alone solar pump will be provided. The state government will give at-least subsidy of 30%; and the remaining at-most 20% will be provided by the farmer.
- Component C: Individual pump solarization (IPS)
    - » Individual farmers having grid-connected agriculture pump will be supported to solarize

pumps. Solar PV capacity up to two times of pump capacity in kW is allowed under the scheme.

- » The farmer will be able to use the generated solar power to meet the irrigation needs and the excess solar power will be sold to DISCOMs.
- » CFA of 30% of the benchmark cost or the tender cost, whichever is lower, of the solar PV component will be provided. The state government will give at-least subsidy of 30%; and the remaining at-most 40% will be provided by the farmer. Bank finance can be availed by the farmers, so that they have to initially pay only 10% of the cost and remaining up to 30% of the cost as loan.
- » In North–Eastern states, Ladakh, Jammu and Kashmir, Himachal Pradesh, Uttarakhand, Lakshadweep, and Andaman and Nicobar Islands, CFA of 50% of the benchmark cost or the tender cost, whichever is lower, of the solar PV component will be provided. The state government will give a subsidy of at least 30%; and the remaining at-most 20% will be provided by the farmer.

- Component D: Feeder-level Solarization (FLS)
  - » Instead of individual solar pumps the states can solarize the agriculture feeders.
  - » Where agriculture feeders are not separated, loan for feeder separation may be taken from National Bank for Agriculture and Rural Development (NABARD) or PFC/REC. Further, assistance for feeder separation may be availed from the Revamped Distribution Sector Scheme (RDSS) of the Ministry of Power. However, mixed feeders can also be solarized.

- Solar plants of capacity that can cater to the requirement of the agriculture load of the selected feeder can be installed through CAPEX/RESCO mode for a project period of 25 years.
- CFA of 30% on the cost of installation of solar power plant (up to Rs 1.05 crore/MW) will be provided. However, in the North–Eastern states, Ladakh, Jammu and Kashmir, Himachal Pradesh and Uttarakhand, Lakshadweep, and Andaman and Nicobar Islands 50% (up to Rs 1.75 crore/MW) subsidy is available.
- The farmers will get day-time reliable power for irrigation free of cost or at tariff fixed by their respective state.

Achievement as on 31 May 2025

	Sanctioned	Installed
Component A	10,000 MW	603.32 MW
Component B	13,32,324 stand-alone solar pumps	8,28,260 stand-alone solar pumps
Component C	55,073 individual pumps solarization	7,660 individual pumps solarized
	35,08,874 feeder-level solarization	4,68,838 feeder-level solarization

Annexe 4.F.: Rooftop Solar Programme [35]

Prime Minister Modi announced the launch of the Pradhan Mantri Surya Ghar: Muft Bijli Yojana (PMSG: MBY) in early 2024 to provide rooftop solar panels to 10 million households.

These rooftop solar panels would provide up to 300 units of electricity per month. Thirty GW of solar capacity would be created.

Rs 75,021 crore has been sanctioned for the programme.

This has the following components:

Scheme component	Outlay
Central assistance to households	Rs 65,700 crore
Incentives for DISCOMs	Rs 4,950 crore
Model solar villages in each district	Rs 800 crore
Incentives for local bodies	Rs 1,000 crore
Payment security mechanism	Rs 100 crore
Innovative projects	Rs 500 crore
Capacity building	Rs 657 crore
Awareness and outreach	Rs 657 crore
Service charge	Rs 657 crore
Total	Rs 75,021 crore

Progress:

- 54,53,281 applications have been received.
- 11,36,856 applications have been completed.*

* Details available at <https://pmsuryaghar.gov.in/#/state-ut-wise-progress>

Annexe 4.G.: Cost Implications of the End of Thermal Power

Revenue from coal mining and thermal power

Revenue stream	Amount (in Rs crore)
Royalty	16,097.29
District Mineral Foundation (DMF) Contributions	4,817.71
Coal Cess	1,753.09
GST Compensation Cess	29,962.63
Total	**52,630.72**

Dividend income from public sector enterprises

Company	Amount (Rs crore)
Coal India Limited	10,000
NTPC Limited	7,000
Total	**17,000**

- Total revenue: Rs 81,000 crore
- Huge potential from reuse of land
- Land held by coal companies: 2.68 lakh hectare
- Assume price per hectare of Rs 4 lakh Estimated land value: Rs10.72 lakh crore
- Land values appreciate rapidly with change in land use and development

CHAPTER 5

Entering the Hydrogen Economy

Introduction

Hydrogen is produced by splitting water molecules through electrolysis. This hydrogen becomes green when zero-emission electricity from renewables is used for its production. Electricity generation from nuclear power plants does not result in carbon emissions and hence could also be used to produce hydrogen which should be considered green.

Hydrogen can also be produced from biowaste. This hydrogen should also be considered green as biowaste is a renewable resource and if the conversion process uses renewable energy, then there is no carbon emission in the production of hydrogen. India is one of the largest producers of biowaste. This has the potential of providing substantial quantities of green hydrogen. The production could take place at dispersed locations all over India. However, a production plant would need assured supply of biowaste. This would be challenging. The production cost of this hydrogen is competitive. It could be delivered to dispersed consumption locations at a lower cost.

Hydrogen is a stored source of energy like fossil fuels. It can be used as a substitute for fossil fuels. It can be stored and transported like gas. It, however, needs special infrastructure and safety measures as it is highly inflammable. There are the hard-to-abate sectors where the complete substitution of fossil fuels by electricity is technically not feasible. When all electricity production becomes free of carbon emissions, these sectors will continue to use fossil fuels and emit carbon dioxide. The substitution of fossil fuels by green hydrogen is being seen as the feasible way of eliminating carbon emissions from the hard-to-abate sectors. Currently, the cost of hydrogen produced through electrolysis ranges from \$4.5/kg to \$6/kg.[1] The cost of electricity from renewables has been coming down. Electrolysers are becoming more efficient. They are also becoming cheaper. Hence, the cost of green hydrogen is expected to come down.

India's Green Hydrogen Mission

India, like all others, would need to eliminate carbon emissions from the hard-to-abate sectors to become net zero. This would be achievable only by entering into a green hydrogen economy. In a strategic forward-looking move, India chose to become an early mover with the ambition to be on the global frontier in the hydrogen economy, along with the advanced industrial economies. This would mean the creation of globally competitive domestic capacities for (a) the production of green hydrogen and (b) the use of green hydrogen to substitute fossil fuels in each of the hard-to-abate sectors.

In his Independence Day address to the nation on 15 August 2021, Prime Minister Modi announced the setting up of a National Hydrogen Mission.[2] The National Green Hydrogen Policy has a comprehensive implementation strategy (Annexe 5.A.). It envisages the production of at least five million tonnes per annum of green hydrogen by 2030, with the potential to reach ten million tonnes per annum with growth of export markets. The need for a whole of government approach has been recognized. Accordingly, an Empowered Group under the Cabinet Secretary has been set up for coordinated decision-making.

The government has already committed Rs 19,744 crore:

- Rs17,490 crore for the Strategic Interventions for Green Hydrogen Transition (SIGHT) programme
- Rs 1,466 crore for pilot projects
- Rs 400 crore for R&D
- Rs 388 crore towards other Mission components

Many state governments have been keen to make their states hubs of the new hydrogen economy and have announced incentives (Annexe 5.B.).

Why is there the need for a government-led Hydrogen Mission? Green hydrogen is considerably more expensive as a source of energy than the fossil fuels it could replace to eliminate carbon emissions. There is, therefore, no demand for green hydrogen. Market forces left to themselves are unlikely to create demand for green hydrogen in the foreseeable future. At the present juncture it is only the government that can create the market for green hydrogen and make its production globally competitive by steering market forces to cut costs with scale, competition, and innovation.

Strategy

Production of green hydrogen is not an end in itself. It is the means to an end. The end is its downstream use to replace fossil fuels from where carbon emissions would get eliminated as a result. There is no demand in the market for green hydrogen for a downstream green product in India. Hence, subsidies for hydrogen production alone may not yield the desired outcomes. The state would need to assume greater responsibility and a more direct role in trying to achieve success in the hydrogen economy. India has an advantage in being an early mover. As a result, the probability of becoming globally competitive is higher. But the government would need to be creative in crafting the strategy and the policies that would deliver, and that too at costs that are affordable. Some new ideas on the way forward are suggested in this chapter.

The following considerations would be important as the journey into the hydrogen economy proceeds:

- Financial resources with the government for providing subsidies would remain a major constraint for the foreseeable future.
- To the extent explicit subsidies from the budget can be avoided or minimized, it would be better.
- Implementing pilot downstream projects at the earliest in every segment would lead to the creation of demand for green hydrogen.
- Repeated bids for procurement can be used to create a competitive industry structure and drive innovation and cost reduction.
- Given India's potential cost advantages in the availability of abundant cheap renewable energy and in project execution, its costs of producing green hydrogen could become among the lowest in the world. This has been demonstrated in the solar power sector. Solar power serves as a major input cost for producing green

hydrogen using electrolysers. Solar power in India is one of the cheapest in the world. Hence, India becoming a cost-competitive producer of green hydrogen is a feasible outcome.

- It would be advantageous to produce hydrogen from biowaste in parallel to take this technology to commercial-scale production and create competitive capacities.
- Becoming the least-cost producers in the world of green hydrogen and then of downstream uses for carbon-free production of goods and services is the outcome the Hydrogen Mission could deliver.
- Focused well-funded technology development and its deployment in the hard-to-abate sectors is the way forward. Technology development through public funds combined with the power of competition to get firms to innovate and reduce costs would be essential.
- Competition and innovation would gather momentum through government-driven procurement and a gradual increase in the size of the market. Increasing market size leads to economies of scale and cost reduction.
- The role of the government in creating demand would decline as private firms begin to generate demand in order to become compliant with the requirement of Carbon Border Adjustment Mechanism in the European market.

R&D and Technology Development

Funding of R&D by the Mission with a narrow focus leading to potential deployment is likely to give better results. The Defense Advanced Research Project Agency (DARPA) of the US is a good example to emulate. Proposals for a well-defined outcome could be invited. These proposals could come from an institution, or a firm, or a consortium. In some cases where the problem appears daunting and the benefits from success appear large, two groups may be funded simultaneously to solve the same problem. This may appear expensive, but the success in reducing costs would far outweigh the higher expenditure of parallel efforts in technology development. Then, unsolicited proposals should also be entertained and funded.

Domain experts should be inducted into the decision-making committee(s) for approving technology development projects. Eminent NRIs may be glad to serve on these committees as they would be able to attend online. These technology development projects should be well funded. It would be prudent to fund fewer projects but funding them fully to reach the actual deployment stage rather than spreading resources thinly. Also, there should be an initial acceptance that many such funded projects may not succeed, and this may be explicitly recorded in the approval of the programme. Apprehensions of time and cost overruns or of lack of success should not inhibit decision-makers. The aim should be to get better, more efficient, and cheaper materials and processes.

Special-purpose Vehicle for Green Hydrogen Procurement

As of now there is no market in the country for green hydrogen. This is the consequence of there being no market for goods and services that would use green hydrogen to replace carbon-emitting fossil fuels and be more expensive.

Conventional support mechanisms, like the Strategic Interventions for Green Hydrogen Transition programme,[3] or the Production-Linked Incentive (PLI) Scheme, may not deliver the desired outcomes in the absence of a market for green hydrogen and products made from it.

It is necessary to understand an investor's perspective. He needs to see sufficient sustained demand in a price range where he can get adequate returns on his investment. On this basis, he is able to raise equity and debt whose providers do their own due diligence. A long-term purchase agreement at a price, the L1 bid price, which generates reasonable returns gives the requisite confidence to financial markets for the provision of both equity and debt. This has been seen in the Solar Mission where all the capacity has been created by the private sector and where financing has not been a constraint.

Procurement of green hydrogen from a prospective date for meeting the downstream demand for different projects whose development is underway would create production capacities. Inviting bids for the supply of hydrogen for each downstream project would facilitate movement down the cost curve and nurture a competitive industry structure. Competitive procurement of green hydrogen for different downstream pilot projects in the quantities and from the dates of the commissioning of these plants is being suggested to create the market for green hydrogen through individual long-term procurement contracts. This could be done more efficiently by creating a special-purpose vehicle (SPV) for such procurement. This SPV could be set up along the lines of Solar Energy Corporation of India (SECI) which through competitive procurement over the years has enabled India to have the cheapest solar power.

The bulk of the procurement by the SPV should be for hydrogen production through electrolysers and for meeting the firm demand of downstream projects. Many firms are positioning themselves to produce green hydrogen using electrolysers. In parallel, electrolyser production capacities are being developed. The competitive process should follow the approach of the National Solar Mission, which in its initial phase successfully created a competitive industry structure. There were repeated rounds of bidding for quantities which were small and then went up. Many developers have emerged in this new area and some of them are large firms now. Competition drove prices down.

Further, developing two pilot plants for producing green hydrogen from biowaste would be a good beginning. One, for establishing an oxy-steam gasification demonstration plant from solid biomass like wood, bamboo, and coconut shells. The other one, a biochemical reactor plant to use urban sewage. Replication with the use of cow dung would be the next phase (Annexe 5.C.). Sites may be chosen. Bids may be invited for the setting up of the smallest techno economically feasible plants. The reliable supply of biowaste for the plant may be arranged along with the price. The characteristics of the biowaste may be given to the bidders who would bid on the price at which they would supply hydrogen. By taking care of the supply of biowaste and its price, the risk for the bidders would be reduced. The bid prices would, therefore, be lower. A longer time may be given for submitting bids to make it easier for Indian firms to tie up with potential technology partners in the world.

Nuclear Power Corporation of India Limited (NPCIL) should also initiate a small pilot project of the minimum technologically feasible size, either from an existing plant or one of the new plants under construction, to try and produce hydrogen by directly splitting the water molecule using the very high temperature heat that is generated in a nuclear reactor. In theory, this is an attractive idea as energy would be used to directly split the water molecules rather than first using it to produce electricity and then using that electricity to electrolyze water to get hydrogen. There should also be assured offtake of this hydrogen for downstream applications to ensure the project's viability.

It would be desirable to create separate SPVs for each industrial segment for the development of pilot plants. These SPVs may be promoted by the ministries dealing with the industries chosen for developing pilot projects.

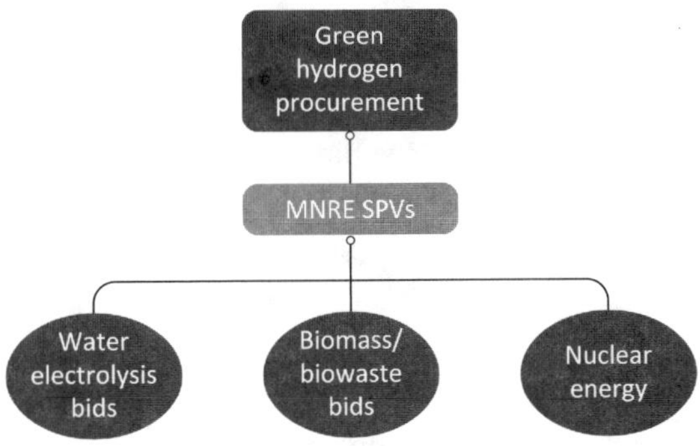

Figure 1 SPV for green hydrogen procurement

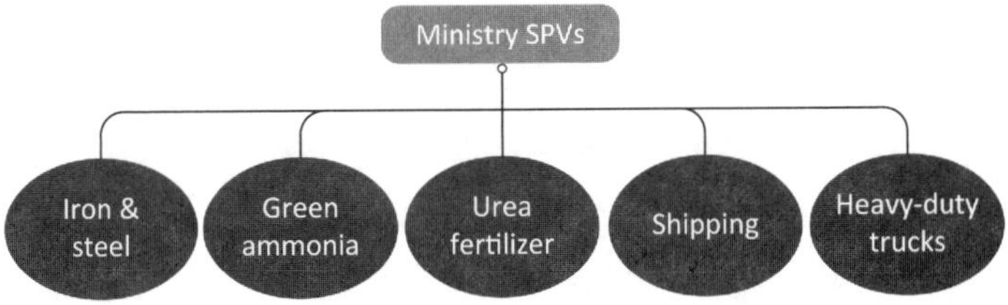

Figure 2 SPVs for downstream green hydrogen-based projects

Where the technology for such a plant is mature with at least one plant in successful operation, the best way to proceed would be to invite bids for the setting up such a plant in India. The bid parameter should be the price of the carbon-free product. Purchase of the entire output of the plant through a long-term procurement contract should be assured. This would de-risk the investment for setting up the plant. The supply of green hydrogen should be assured along with its price for the first few plants. The supply of competitively procured green hydrogen would lower the cost of downstream production. Further, it could also reduce the risks of potential bidders as they would only need to set up the plant, run it, and not have to ensure the supply of green hydrogen for their plant. This would change once a competitive market for the supply of green hydrogen emerges in the coming years.

This should get competitive bids. It would be necessary to announce the intention of inviting bids so that potential bidders get enough time to prepare. After the shortlisting of bidders, the bid and contract documents may be finalized transparently in consultation with the short-listed bidders. This approach was tried in the ultra-mega power projects (UMPPs) where the winning bid price was far below the most optimistic expectations. These bids may be the first of their kind in the world. Getting the bid process right would be essential for success.

One pilot plant using green hydrogen having zero emissions in each downstream industrial segment should be developed. The work for developing such plants should be initiated simultaneously. These industrial plants should be of the minimum technically feasible size. This would keep the capital investment for a new green plant on the lower side. The cost of using the more expensive output would also be lower.

After price discovery it would be easier for the government to see how much of a transition by way of new green capacities is affordable and sustainable. It can then determine inter se priorities on a more rational basis. It would also determine the pace at which bidding for new plants would be undertaken, keeping in mind the reduction in carbon emissions, the additional cost of the carbon-free product, and the ability of the system to absorb this extra cost. The extra cost of the output from the pilot plants would have to be absorbed by the government through tax concessions and subsidies.

Iron and Steel

The technology for a fully carbon-free steel plant is at a nascent stage. Setting up a greenfield plant to produce carbon-free steel would put India in a global leadership position. The risks of technology, time and cost uncertainties in such a new project are too high for private capital in India. Such a plant can be set up if the government puts in the money and takes all the risk. The government can derisk its investment fully by deciding to buy all the steel produced on a cost-plus basis. This steel would be more expensive than normal steel. The government could use this steel in its own projects. For these projects, the additional cost of expensive steel would have a marginal impact on the final cost of these projects which the government agencies can comfortably absorb.

The project should be implemented by a dedicated special-purpose vehicle with all the major steel firms being partners. They should be required to put in some equity in order to have a stake in success. The government could provide equity through its Infrastructure and Investment Fund and guarantee debt. Direct financing from the budget would not be needed. The management and staffing should be joint, in a genuine partnership mode. There would be joint learning in absorption of the technology, project execution, and smooth running of the plant. Once this plant begins to work smoothly, the actual higher cost of this steel would be discovered. The partnering steel producers would have the ability to build such carbon-free steel plants on their own. Project execution by a consortium of steel firms, fostering collaboration and shared capacity development on this scale may be a first in India and would be a useful precedent.

Then the partners can decide when the international market will make it worth their while to build such plants. The government would be in a better position to take a view on the pace at which it would like to get such plants built in India for the domestic market and the policy instruments it would consider appropriate.

Steel plants based on direct-reduced iron process using natural gas or syngas from coal have the advantage that it is technically feasible for them to be retrofitted to use green hydrogen instead of natural gas or syngas (Annexe 5.D.). Emissions would then decline by 95% with green electricity integration.[4] The government could provide fiscal incentives for new steel plants that would use the direct-reduced iron process. The incentive could be through a lower Goods and Services Tax (GST) rate. Firms opting for such steel plants could switch over to the use of green hydrogen at a modest cost of retrofitting and produce steel that would have an export market in Europe and elsewhere as the standards for embedded carbon become more stringent under Carbon Border Adjustment Mechanism. At some time in the future, the government could consider mandating that all new steel plants be built using the gas-based direct-reduced iron process, when it feels that the higher cost of steel from these plants can be absorbed by the Indian economy. The cost of steel in the final goods for which it is used is in any case quite modest.[5,6] A level playing field vis a vis existing older plants can be provided by imposing a carbon tax on the steel produced by them. Import-duty protection may then be given to domestic producers against cheaper imports, a Carbon Border Adjustment Mechanism type of import duty. India could become a global leader in the production of green steel and its export.

Fertilizer

A pilot project for producing carbon-free urea would need a distinct approach. The production cost per tonne of green urea fertilizer would be higher. This could be one of the first plants in the world. The key change would be the use of green hydrogen for the production of ammonia instead of natural gas (Annexe 5.E.). Green ammonia needs carbon dioxide as an input for the production of urea. This carbon dioxide would have to be provided by carbon capture. Ideally, this should come from cement plants. There is no technology as yet for eliminating emissions fully from cement production. Carbon capture and use is therefore a convenient solution for achieving zero emissions from cement production. The cost of green urea would be considerably higher due to the cost of captured carbon dioxide and the higher cost of green ammonia.

New locations of cement and fertilizer plants should be strategically planned to ensure they are in proximity, enabling a constant supply of CO_2 from a cement plant, with minimal transportation costs, for urea production from green ammonia. Bids may be invited for carbon capture and supply from a cement plant for the green fertilizer plant. This could also be one of the first such large-scale plants for carbon capture and use.

Bids may then be invited for the setting up of the pilot urea fertilizer plant. The bid process including the bid and contract documents may be evolved in consultation with prospective bidders. The supply of green hydrogen and CO_2 may be assured along with their prices as part of the bid process. The bid parameter in this bid would be the cost of converting green ammonia and captured carbon dioxide into green urea.

The government would have to buy this fertilizer and give it to farmers at the price at which it is giving them normal fertilizer. For this green fertilizer, there would be a much higher subsidy burden on the government which it would have to absorb. After discovering the cost, the government could then take a rational view on the pace at which new non-carbon dioxide-emitting plants for cement and urea would be built.

The key trade-offs to be considered by the government would be:

- Movement down the cost curve by getting more zero-emission plants built through competitive bidding. Costs decline as the size of the market grows. Competitive bidding incentivizes cost reduction.

- The greater the quantity of green urea being procured by the government, the higher would be the subsidy burden on the government as this green urea would be given to the farmers at the same price as other urea.
- To get to net zero, there would be the cost of the closure of the new carbon-emitting normal fertilizer and cement plants that would be built in the coming years.
- This cost would be lower to the extent the share of these plants in new plants becomes lower. On the other hand, the subsidy burden on the government for buying green urea would rise. An estimated net present value (NPV) cost comparison on an ongoing basis would help decision-making.

Heavy-duty Trucks

Green hydrogen is an option for moving heavy-duty trucks without carbon emissions. Batteries for heavy-duty trucks are yet to become feasible. Initially picking up a few busy routes for implementing pilot projects would be the optimal way to proceed. The capital cost of a heavy truck would be higher than that of a normal truck. Hydrogen-filling stations would need to be put up on the route with the assured supply of green hydrogen. Per kilometre cost of using green hydrogen as a fuel would be higher. Both the truck and the hydrogen would need to be subsidized. A critical mass by way of trucks and hydrogen-filling stations has to come up simultaneously. This can be best done through a state SPV with the backing of the state government. Announcements of subsidies for trucks and hydrogen-filling stations may not generate the market response to achieve the critical minimum size of operations needed.

One way of implementing the pilot project would be for the SPV to invite bids for a large number of trucks to have the volumes for getting a competitive price. These trucks may then be leased out at a subsidized rate, which makes their use appear attractive to truck operators. For the supply of hydrogen, the competitive procurement of green hydrogen from biowaste may be tried out. This would have two advantages. First, green hydrogen plants using biowaste would come up. It would be easier to get such plants developed for transport rather than for industrial applications. Further, there would be lower costs in the transport of the green hydrogen being produced in smaller units along the highway. For the setting up and running of the filling stations, the approach being used for petrol and gas-filling stations should work. The SPV should see that there is enough volume of business for these stations and normal profit margins.

For the selected routes a holistic plan would have to be prepared by the SPV of the number of trucks, hydrogen-filling stations and the volume of green hydrogen to be procured for these stations. Again, the sale price of green hydrogen would have to be subsidized by giving the truck operator an attractive per kilometre cost vis a vis a normal diesel truck. The subsidies could be given by lowering the Goods and Services Tax rates for the hydrogen trucks as has been done for electric vehicles and, if necessary, this may have to be topped up by cash subsidies. After the learning experience from the initial pilot projects, decisions on the pace of scaling up would become easier.

Shipping

Green ammonia can be used in shipping. The use of green hydrogen directly or through green ammonia for decarbonizing shipping are options. India could do pilot projects for these options. Price discovery by provision of green hydrogen and green ammonia at actual cost would be the way to go forward.

Kerala, for instance, could set a target date when all vessels being used for tourism would be carbon free. This would enhance the brand value of Kerala tourism. This can be done by the state facilitating the transition without any financial burden as the higher cost would be passed on to the tourists. Further, some coastal green shipping pilot projects can be implemented in a similar manner.

From a long-term perspective, India needs to create facilities for the supply of green hydrogen/green ammonia at its ports to meet the demand of international shipping as it moves away from fossil fuels to become net zero. New engines using green hydrogen and green ammonia for shipping are needed for the global transition to zero emissions from shipping. India could take a strategic view and try to develop these engines with public money in a public–private partnership mode. To the extent it succeeds, it could become a competitive hub for ship building.

Power generation

The use of green hydrogen and green ammonia may also be tried out for power generation to discover the cheaper and better option. When the full decarbonization of electricity is undertaken, one of these green fuels would be needed to run a normal plower plant for meeting the seasonal peaks in demand. Pump storage projects (PSPs), concentrated solar projects (CSPs), and battery energy storage systems (BESSs), all offer only short-term, usually daily, storage for getting carbon-free electricity round the clock.

Future Journey

As the first set of downstream pilot projects get completed and actual costs get discovered, the cost of a tonne of carbon mitigation in a zero-emission process in different industries and processes would be discovered. This could begin happening around the end of this decade if decisions are taken by the government to set up the SPVs for the development of these plants, and these SPVs begin the setting up of the pilot downstream zero-carbon-emitting plants immediately.

A rational view on prioritization and scaling up would then be possible. On scaling up, the higher cost of the green carbon-free product would have to be borne by the consumer and/or the government. Decisions would emerge from the assessments of the absorptive capacities of the consumer and the government. For the first few plants, the higher costs would have to be borne by the government. It can do so by providing lower Goods and Services Tax rates and, if necessary, by also giving cash subsidies. The template for such support has been successfully used to promote electric vehicles. Once the market has grown and there is competitive production capacity, the government could take the view that the higher cost can be passed on to the consumer partially, and then fully. This can be done by first imposing a carbon tax on older existing production capacities and using these revenues to partially meet the cost of subsidizing new green plants. Then the subsidies for the green plants can be phased down and the level playing field maintained by increasing the carbon tax. Regulation mandating that all new plants have to be carbon free from a prospective year becomes feasible, once there is a level playing field between new carbon-free plants and the old ones using fossil fuels. Subsidies for green production can be phased out altogether by raising the carbon tax to maintain the level playing field. An equivalent tax may be imposed on non-carbon-free imports to maintain a level playing field. This would be an Indian version of Carbon Border Adjustment Mechanism. These decisions would be challenging. A political call would have to be taken on the financial support that the government can provide and the extent to which the consumer can be made to pay a higher price.

Regulation mandating that new capacity should be carbon free from a prospective year would result in peaking of carbon emissions in that sector. This would be a major inflection point. After this, the decision on the pace of retirement of existing capacities would be relatively easier. Plants with the highest carbon emissions per unit output should retire first. The cost of premature retirement would be borne by the investors who in any case would have earned adequate returns by then. The movement from peaking to net zero could be faster. It must be borne in mind that the earlier the peaking takes place the lower would be the costs of the transition to net zero.

It would be prudent to use scarce resources for accelerating green downstream production to get a greater share of export markets rather than subsiding export of green hydrogen. An important new dimension is the European Union's (EU) Carbon Border Adjustment Mechanism.[7] The export market for green products which have a higher cost is being created. India could potentially leapfrog and become a major producer and exporter of green carbon-free manufactured goods.

For this, the government could create a dedicated development financial institution (DFI) which provides long-term credit at low-real interest rates for the creation of new manufacturing capacities which meet the new European Union carbon standards. These new plants would be for exports while the existing plants could go on supplying to the domestic market. With this ring fencing, competitive capacities would emerge and movement down the cost curve would take place without any additional domestic costs being incurred. Forward-looking Indian firms with the help of soft financing could acquire economies of scale and benefit from agglomeration effects. They could become globally competitive in the new green hydrogen economy.

India could become a green manufacturing hub. The costs of the transition into the hydrogen economy would have been well worth it. India could get the success in the export of manufactured goods that has been eluding it so far. In the process, the aspiration of becoming a developed industrial economy would also be fulfilled. The pursuit of the goal of becoming net zero would help in the transformation into a developed industrial economy.

Annexes

Annexe 5.A.: National Green Hydrogen Mission

The National Green Hydrogen Mission (NGHM), launched by the Government of India in January 2023, aims to make India a global hub for the production, use, and export of green hydrogen and its derivatives.

Mission objectives

- Establish 5 million metric tonnes (MMT) per annum of green hydrogen production capacity by 2030, with potential to scale to 10 MMT.
- Enable associated renewable energy capacity addition of ~125 GW.
- Create over 6 lakh jobs, attract Rs 8 lakh crore in investments, and avoid 50 MMT/year of CO_2 emissions.
- Promote indigenous manufacturing of electrolysers and green hydrogen technologies.
- Phase I (2022–26)
- Focus on demand creation in existing hydrogen-using sectors (refineries, fertilizers).
- Expand domestic electrolyser manufacturing.
- Pilot projects in steel, heavy mobility, shipping.

Expected impact by 2030

Indicator	Target
Green hydrogen capacity	5 MMT/year
Renewable energy addition	125 GW
Jobs created	6,00,000+
CO_2 emission reduction	50 MMT/year
Investment attracted	Rs 8 lakh crore

Annexe 5.B.: State-level Policies and Initiatives

The Green Hydrogen Mission has fortunately received enthusiastic support from many states who have come up with their own policies and programmes. The major ones are:

Andhra Pradesh Green Hydrogen and Green Ammonia Policy

- Target: 0.5 million tonnes per annum (MTPA) green hydrogen production capacity or 2 MTPA green ammonia production.
- Incentives for green hydrogen projects:
  - » Reimbursement of net State Goods and Services Tax (SGST) revenue from sale.
  - » 100% exemption of electricity duty for the power consumed.
  - » 25% of intra-state transmission charges shall be reimbursed for a period of five years.
  - » Allocation of government land for development on priority basis at lease rate of Rs 31,000 per acre per year.
  - » 100% exemption from payment of land-use conversion charges and from payment of stamp duty.

Gujarat Hydrogen Policy

- The Gujarat 'Policy 2023', gives priority to leasing land for the development of green hydrogen production within the state. The lease period for solar, wind, and wind–solar hybrid projects dedicated to green hydrogen production is set at 40 years. Gujarat is providing leading energy companies such as Reliance and Adani with land required for producing green hydrogen and generating the solar power needed to produce it.

Haryana Hydrogen Policy

- Haryana Hydrogen Policy aims at a production target of 250 kilotonnes per annum (kTPA) green hydrogen by 2030, and electrolyser manufacturing capacity of 2 GW.

Karnataka Hydrogen Policy

- In Karnataka, a project aims to establish a 1.2 MTPA green hydrogen and green ammonia plant alongside a captive solar power unit.

Kerala Hydrogen Policy

- The initiatives include the announcement of a Rs 200 crore scheme to set up green hydrogen hubs in Kochi and Thiruvananthapuram. Kochi Green Hydrogen Valley project aims to establish green hydrogen and ammonia plants, along with renewable energy infrastructure. ReNew Power has proposed an investment of Rs 26,400 crore for a green hydrogen project in Kerala, aiming to establish a production capacity of 220 kTPA of green hydrogen.

Madhya Pradesh Hydrogen Policy

- Green hydrogen projects are eligible for investment assistance of up to Rs 2 billion. There are special incentives for elctrolyser manufacturing.

Maharashtra Green Hydrogen Policy

- Production target of 500 kTPA by 2030
- An outlay of over Rs 8,000 crore

- Policy benefits:
  - » Exemption from cross-subsidy surcharge and electricity duty.
  - » 50% exemption from intra-state transmission charges and wheeling charges.
  - » Priority of land allotments.
  - » 1% interest subsidy on hydrogen transport project.
  - » Subsidy of Rs 50 per kg for blending green hydrogen into gas and capital subsidies for the establishment of 20 hydrogen refuelling stations and 500 hydrogen fuel cell-based passenger vehicles.

Punjab Hydrogen Policy

- Punjab Hydrogen Policy targets green hydrogen/ammonia production capacity of 100 kTPA by the year 2030 and establishment of a Centre of Excellence (CoE) with different academic and research institutions and industries.

Rajasthan Green Hydrogen Policy

- Ambitious target: 2000 kTPA energy production by 2030.
- Incentives for projects:
  - » 50% rebate on transmission charges and exemption from surcharges for 10 years.
  - » Priority in land allocation.
  - » 30% subsidy for green hydrogen research centres.
- Waiver of charges:
  - » 100% waiver on wheeling and transmission charges.
  - » Reimbursement/waiver of banking charges for 7–10 years.

- Investor incentive: Exemption in transmission and distribution charges.

Tamil Nadu Hydrogen Policy

- Tamil Nadu is planning to develop a 5,000 MW solar PV plant, a 1.5 GW electrolyser, and a 1.1 million tonne per annum ammonia synthesis loop plants.

Uttar Pradesh Green Hydrogen Policy

- Provides a capital expenditure subsidy of 60% for electrolysers with a minimum capacity of 50 MW.
- Offers monetary incentives (Rs 3,500 per tonne of green urea, 30% one-time grant support for technology purchases up to Rs 50 million), and supports ecosystem creation and skill development.

West Bengal Green Hydrogen Policy

- Establish State Centre of Excellence for R&D.
- 100% waiver on land-use conversion tax, land registration and stamp duty and electricity duty.

Annexe 5.C.: Hydrogen Production Through Biomass and Biowaste

Biomass and biowaste can be converted into green hydrogen through biomass gasification or biogas reforming. At high-temperature, biomass releases synthesis gas or syngas, which includes hydrogen (H_2), carbon monoxide (CO), and other gases. The hydrogen present in syngas can be isolated and purified to produce green hydrogen. The minimum feasible plant size for biohydrogen production is 250 kg per day, with a cost of production around USD 3 per kg of biohydrogen.

Indigenous technology: A 250 kg/day biohydrogen pilot plant has been developed by the Indian Institute of Science, Bangalore. It produces 100 g of hydrogen from 1 kg of biomass. The plant can produce 0.25 tonnes of hydrogen per day.

Annexe 5.D.: Decarbonization of Steel Production

India's steel industry is one of the highest contributors to industrial greenhouse gas (GHG) emissions. There is heavy reliance on coal-based processes such as the blast furnace–basic oxygen furnace (BF–BOF) route and coal-based direct-reduced iron (DRI) making. Transitioning to renewable energy (RE) and green hydrogen (GH_2)-based steelmaking has to take place.

Current steel production routes and emissions

India's steel production is primarily dominated by:

Blast furnace–basic oxygen furnace (BF–BOF)

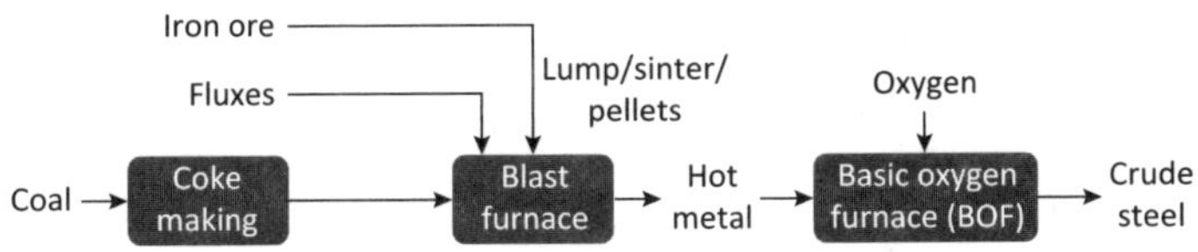

This is the conventional integrated route, using coking coal to convert iron ore into hot metal in the blast furnace, followed by refining in a basic oxygen furnace. This route accounts for over 45% of India's crude steel production and emits approximately 2.2–2.6* tonnes of CO_2 per tonne of crude steel.

Coal-based direct-reduced iron with electric arc furnace

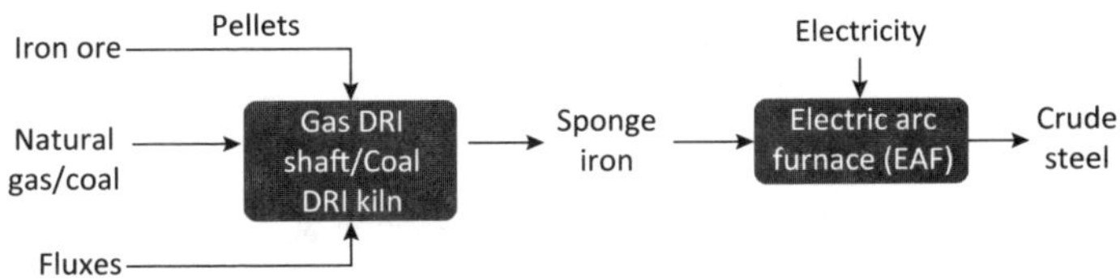

* *Greening the Steel Sector in India: Road Map and Action Plan*. 2024. Ministry of Steel, Government of India

India is the largest producer of coal-based direct-reduced iron, where non-coking coal is used in rotary kilns to reduce iron ore. The resulting sponge iron is then melted in an electric arc furnace. This route emits approximately 2.7–3.1 tonnes of CO_2 per tonne of crude steel and dominates the secondary steel sector.

Gas-based direct-reduced iron with electric arc furnace

Limited in capacity, this route uses natural gas in shaft furnaces (e.g., MIDREX, HYL), with lower carbon intensity of about 1.4–1.6 tCO_2/tcs. This route is mainly present in western India, where natural gas availability is relatively better.

Decarbonization pathways

Two promising pathways for reducing emissions are:

1. Green hydrogen and renewable energy-based direct-reduced iron-electric arc furnace route.

This involves the production of direct-reduced iron using green hydrogen. The subsequent melting of direct-reduced iron takes place in an electric arc furnace powered by carbon-free electricity.

Green hydrogen can replace syngas (a mix of CO and H_2) or natural gas as the reducing agent in shaft furnaces.

This process has near-zero direct emissions. This steel is however more expensive.

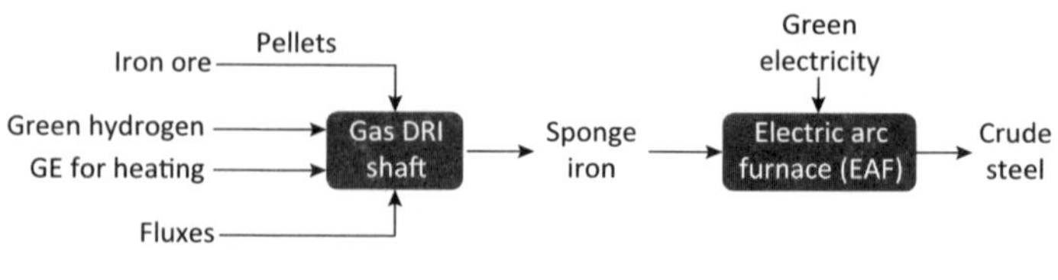

2. Scrap-based electric arc furnace with green electricity

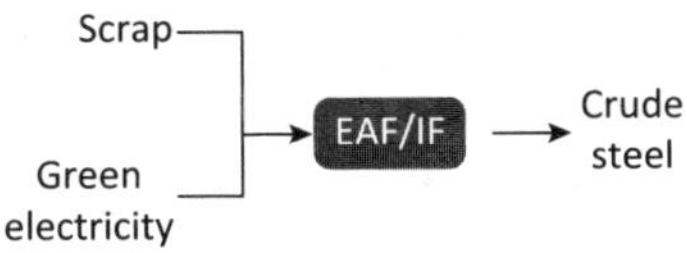

In regions with sufficient scrap availability, recycling scrap in an electric arc furnace powered by green electricity is the most energy- and carbon-efficient pathway. For making steel, this route bypasses the need for iron ore reduction entirely and can achieve zero emissions.

Transitional steps

- Gas-based direct-reduced iron using natural gas or blue hydrogen: Transitioning existing coal-based direct-reduced iron kilns to gas-based shaft furnaces using natural gas can reduce emissions by ~30–40%. These systems can later be upgraded to operate with green hydrogen, enabling a phased transition.
- Hybrid direct-reduced iron units: Modern direct-reduced iron plants that are capable of operating on natural gas and later shifting to green hydrogen offer a viable bridge solution.

Annexe 5.E.: Fertilizer Industry

Urea is the most widely used nitrogen-based fertilizer in India. The conventional production process of urea is highly carbon intensive. The process typically involves reforming natural gas to produce hydrogen, which is combined with nitrogen from air to synthesize ammonia (NH_3). This is followed by its reaction with CO_2 to produce urea ($CO(NH_2)_2$).

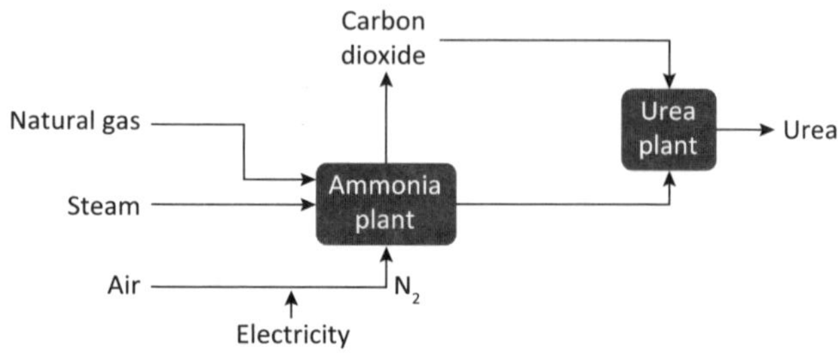

Emissions are produced from fossil fuel combustion for hydrogen production and energy use in air separation and synthesis units.

Decarbonized pathways for urea production

Decarbonization of urea production route requires the use of green feedstocks and clean energy inputs at all major steps:

- Green electricity for nitrogen separation: Green electricity can replace conventional electricity in air-separation units (ASUs) for extraction of nitrogen (N_2) from air.
- Green hydrogen for ammonia synthesis: The existing natural gas steam-reforming process can be replaced with green hydrogen, produced from electrolysis of water which would use renewable energy. Green hydrogen thus produced combines with nitrogen to synthesize green ammonia (NH_3) with near-zero direct emissions.

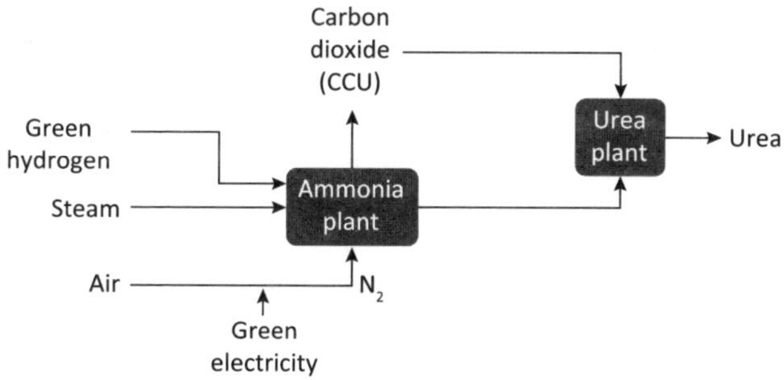

- External CO_2 utilization for urea formation: In place of existing fossil fuel-based reformers, CO_2 can be sourced from CO_2 captured from industries such as cement produced in large quantities. This would be a win-win situation for both industries.

Alternative route: biomass gasification and renewable energy electricity integration

Another decarbonized pathway involves biomass gasification. Syngas (H_2 + CO) derived from sustainable biomass can replace fossil fuels as the feedstock for hydrogen production. The hydrogen-rich syngas can be cleaned and made into ammonia, with green electricity used for air separation for nitrogen and the productive process. This approach is carbon neutral.

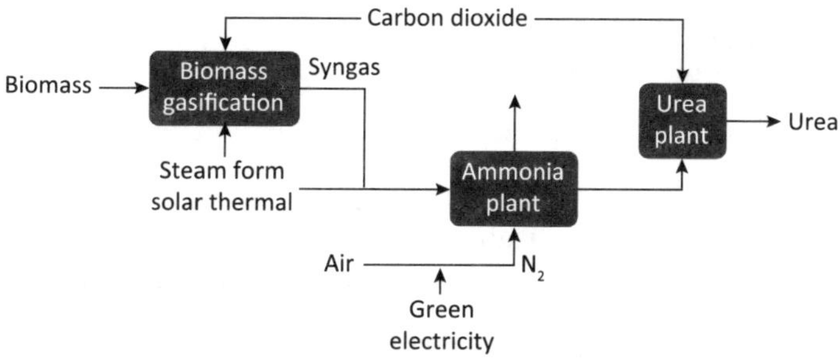

Annexe 5.F.: Cement Sector

The cement industry is one of the most carbon-intensive sectors.

Present process

Cement production involves extracting and processing of raw materials (limestone and clay) which are heated in a rotary kiln at high temperatures (1400°C) to form clinker. The clinker is then ground with gypsum and other supplementary cementitious materials (SCMs) to produce different cement types. Coal and petcoke are the predominant fuels used for the thermal energy of about 95% of the production process.

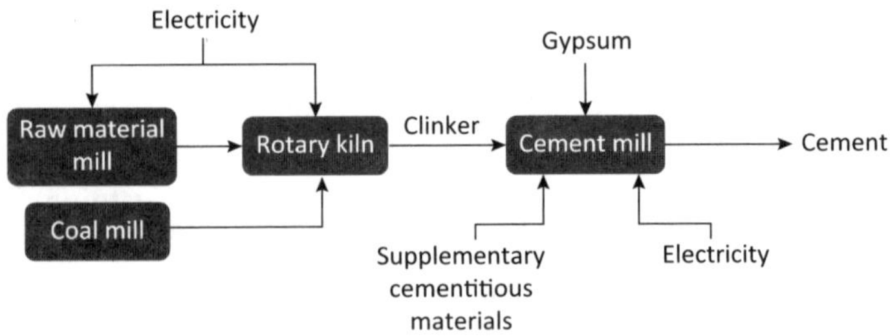

Decarbonization pathway

Shift to low-carbon fuels: Shift from fossil fuels to low-carbon fuels like green hydrogen, biomass, and societal waste like municipal solid waste (MSW) and industry waste would help in minimizing Scope-1 emissions.

There is a potential for kiln electrification which needs R&D efforts.

Carbon capture, utilization, and storage (CCUS): Cement industry would require carbon capture technology options to achieve net zero. The captured CO_2 can be used for fertilizer production and other uses. Indicative cost of carbon capture is estimated at USD 45–90 per tonne of CO_2, while CO_2 transportation and storage add about US 110 per tonne. These costs are high.

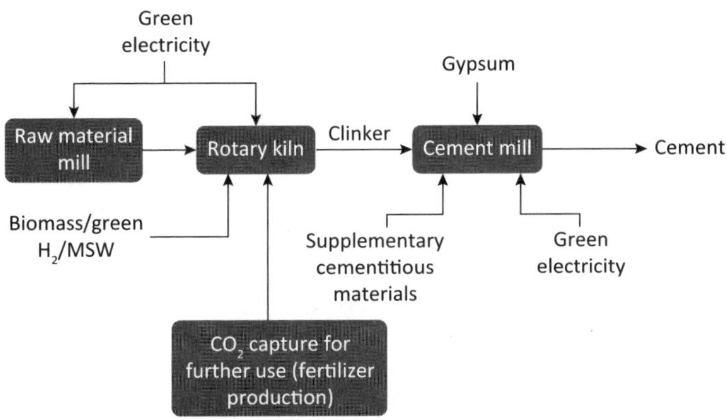

CHAPTER 6

Journey to Net Zero

Introduction

In an act of responsible climate leadership, India declared at COP26 in Glasgow in 2021[1] that it would become net zero by 2070. The declaration of the G20 in 2024[2] in Delhi affirmed that:

> "We reiterate our commitment to achieve global GHG emissions/carbon neutrality by or around mid-century..."

The Trump Presidency is, hopefully, a temporary aberration. Humankind has no option but to move vigorously to achieve net zero at the earliest. This is the most important global 'public good'. The survival of human civilization depends on this. With one-sixth of humanity, India's success in becoming net zero would be critical. This book argues that India can and should succeed.

India is a developing country which aims to eventually become a developed economy. As its economy grows, its per capita income and energy consumption would keep rising. It would not be unrealistic to assume that these would reach, say, current EU levels at some time in the future. India has a long journey ahead in becoming a developed economy. How does one think of the journey to net zero and will it constrain India's development goals? Does it make sense to focus on net zero now? It certainly does. India's transition to a net-zero economy has to be such that it does not have an adverse impact on economic growth and development. This would be possible only with the decoupling of growth from increase in carbon emissions. The trajectory to net zero would be as follows:

This aggregate trajectory would be the sum of the different trajectories of each major carbon-emitting sector.

The emissions would peak in a sector only when all additional production takes place without emissions. Decoupling of growth from an increase in carbon emissions would follow. Output and energy consumption would keep rising, but carbon emissions would not. This would be the critical inflection point for each sector. Once the peak is reached, emissions will plateau. This would mean that emissions would remain at their peak level till the time the existing carbon-emitting sources continue to function. This peak may remain at a plateau for a shorter or a longer time period. Emissions would start declining when the carbon-emitting plants begin to get closed.

This transition path would be specific for each sector. As this imposes a higher cost, market forces on their own would not take the path to this transition. There is, therefore, no alternative to the government assuming direct responsibility of getting the economy to move towards net zero. For this it would need to promote technological development, deployment of frontier, cutting-edge technologies, and cost reduction. These in turn would be the key factors in shaping government policy and the pace of progress. The energy transition to net zero would have to be steered by the government for each carbon-emitting sector. Its policy choices would be decisive in shaping outcomes and costs at each stage of the journey.

Trajectory Framework

This trajectory to net zero can be seen in a simple stylized way. Emissions would rise and then peak. Peaking can occur earlier at one level of total emissions, P1, or later at a much higher level, P2. In Figure 1, P2 implies, total emissions are much higher when peaking occurs and there is a much larger carbon-emitting production capacity. New additional production capacity that is being created after peaking has zero emissions. Output grows but not emissions. Government policy will be driving the trajectory, whether it is P1 or P2. The government would have to assess the relative costs of the alternative pathways and take a political call.

The peak could remain for a shorter or a longer duration. Decline would begin when carbon-emitting plants start getting shut down. This has a cost. The process of closure

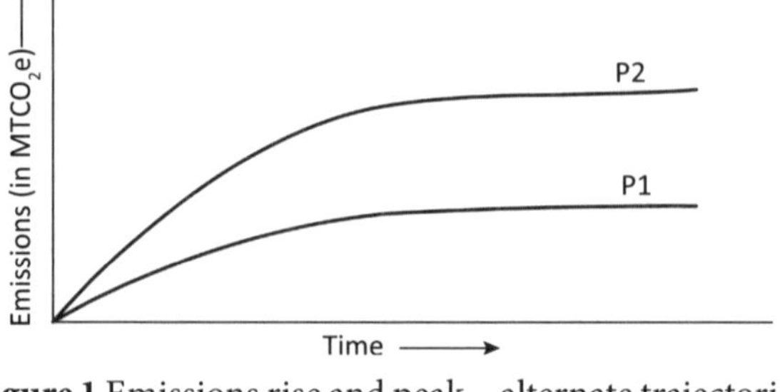

Figure 1 Emissions rise and peak—alternate trajectories

would begin with a government policy mandate. This decline could be gradual, as shown in path D2, or rapid, as in D1 depending upon the rate of closure of carbon-emitting plants (Figure 2).

Government policy would determine the pace of closure, whether the decline trajectory follows D1 or D2. Policy decisions would have to consider the costs of closure. The cost of P1 or P2 on the one hand and D1 or D2 on the other must be assessed and the least final cost pathway chosen. This entails difficult choices for policymakers, made more difficult as these costs are themselves uncertain and would be evolving over time. Prima facie, peaking at a lower level as in P1 should result in a lower final cost of becoming net zero as the carbon-emitting plants which have to be shut down are fewer.

The trajectories, and their costs, would vary from sector to sector. The end goal would be to get emissions to decline and go to zero, while production, energy consumption, and GDP continue to rise. This is the conceptual framework within which analysis of the transition is useful.

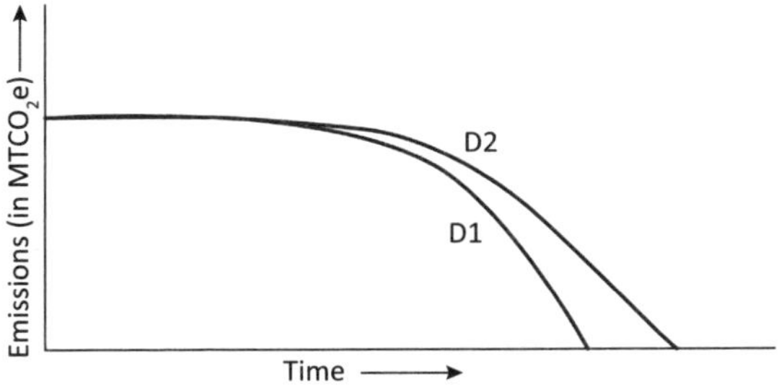

Figure 2 Emissions decline—alternate trajectories

This journey to net zero can be envisaged in three parallel broad pathways. These are:

1. Transforming electricity generation to have zero emissions. This is a major achievable and affordable milestone. This has been analysed in Chapter 4.
2. Electrifying as many sectors of economic activity as possible; some fully and some partially. This has a cost which would vary across sectors. But it would in most cases be affordable. Electrification would make a sector's emissions a part of the total emissions from electricity generation. The sector's decarbonization would be subsumed within the transition of electricity generation and would become zero when emissions from electricity become zero. There would be no additional costs for the sector over and above the initial cost of electrification.
3. Ending the use of fossil fuels in the sectors which cannot be electrified. This is the most difficult. Technologies are still evolving. Costs are quite high at present.

Broad contours, milestones, and inflection points with implications have been analysed in the earlier chapters. Feasible policy options would need to be evaluated on a continuing basis. Affordable costs and benefits would be the key considerations along this complex and challenging journey.

This is going to be a fundamental transformation across the board of the modern industrial economy as the world has known it. Nothing like this, and on such a vast scale has ever been attempted before. It would be prudent to start planning for net zero now. It is the responsible thing to do, rather than taking the easier and softer option of thinking that 2070 is so far away that it would be premature to think and plan for it right now. Better leave it for later generations of planners. There is, however, the phenomenon of the "late-mover's advantage" which becomes very important when thinking about this transition. India has experienced the late-mover's advantage in some sectors. Most Indians have leapfrogged into using mobile phones without seeing land lines. They are now using the internet with their smartphones, bypassing the use of desktop computers and laptops. Their costs have been so much lower as a result. This potential gain should be a material consideration in shaping policies and decisions on the pathways to net zero. The choice is between imposing a higher cost in the near term by mandating that all new plants should be zero or low-emission plants, depending on the state of development of new green technologies and their costs, or bearing the much higher cost later of closing a larger number of carbon-emitting plants. With the passage of time and the creation of many more of the traditional fossil fuel-based industrial plants, this potential "late-mover's advantage" would keep diminishing.

The decoupling between sectoral growth and increase in carbon emissions is the first critical milestone for any sector. Decoupling would result in the peaking of carbon emissions in a sector. Growth would thereafter continue without having an impact on the journey to net zero. This would be the sweet spot. Further, the cost of closing existing carbon-emitting plants would peak at the same time. There is a less well understood but very important implication of this. The longer it takes to decouple growth from increase in emissions, the larger would be the stock of fossil fuel using carbon-emitting plants, and the higher would be the cost of closing them. These costs would be high and vary from sector to sector. Without their closure, net zero would not be achieved.

Therefore, it would be prudent to factor in the cost of the closure of a plant or facility using fossil fuels by 2070, the target year for becoming net zero, for any new large investment decisions in the pipeline. Coal-based thermal plants built in the early 1980s are still running well. New ones being built now would have years of productive life left in 2070. Similarly, heavy industrial processing plants such as in steel, fertilizers, and chemicals have a long life. But all these fossil fuel using plants would need to be shut down. It may turn out to be the case that choosing to build a non- or a low-fossil fuel using plant in some sector, even if the initial capital cost is higher, may be cheaper in the long run if the cost of premature closure by 2070 is correctly factored in. There may already be a case emerging for the reconsideration of planned investments in new refineries, LNG terminals, and long-term 'take or pay' contracts as the demand for electric vehicles is rising and the demand growth rate may already be declining.

Where was India at the time of taking the decision to become net zero by 2070?

Its total emissions (without LULUCF*) in 2020 were 2,959 million tonnes of carbon dioxide equivalent ($MTCO_2e$). The more relevant parameter is per capita emissions, which were 2.1 tonnes of carbon dioxide equivalent ($T\ CO_2e$). For perspective, the per capita emissions of the US were 18 $T\ CO_2e$, and the global average was 6.3 $T\ CO_2e$.[3]

The sectoral shares of these emissions are shown in Figure 3.

* Land use, land-use change, and forestry

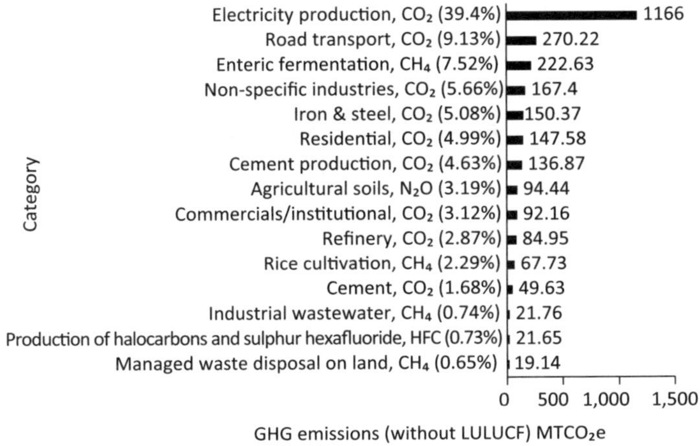

Figure 3 Share of greenhouse gas emissions by category, 2020

Electricity

Electricity generation accounts for the largest share, 39%, of total carbon emissions in India. These emissions are 1,166 $MTCO_2e$.

There is a very recent positive development that it would now be cheaper for the consumer if all additional electricity demand were to be met by renewable energy and sufficient storage of electricity, rather than by building new thermal plants. This has been elaborated in Chapter 4.

When the decision is taken to stop building new thermal plants, carbon emissions would peak within five years as the construction of a new thermal plant usually takes less than five years. Until then emissions would keep rising. They would, thereafter, stay at a plateau. The central and state governments, and the regulatory commissions should decide to give the highest priority for the development of storage along with renewable energy so that they can take the decision that the development of a new thermal plant is no longer necessary to maintain reliable electricity supply. This would save money. Peaking of emissions is a consequence in this case of choosing the least-cost pathway for meeting the increasing demand of electricity.

Emissions would remain at a plateau until the government decides that the retirement of the existing thermal plants should begin. The policy options and trajectories with cost implications have been spelled out in Chapter 4. The costs of closing all thermal plants and coal mines would be large. These would keep getting larger as more plants are built and new mines are developed. Conceptually, the costs of closing can be offset, partially or even fully, by reusing the land which becomes available with the closure of mines and plants. Developed land with infrastructure is needed in large quantities to meet India's huge development needs. This would be very challenging as reuse of land on this scale would be attempted for the first time. Demand for developed land is fortunately not a constraint. Initial success in developing a new township, for instance, leads to increase in prices. There is increasing escalation of land prices thereafter. The extent to which costs of closure would be offset, partially or fully, would depend on the skill and imagination with which redevelopment projects are designed, implemented, and marketed.

The government would have to decide when the retirement of existing plants should begin, and the pace at which plants should be shut down. Not easy decisions. But these decisions may need to be considered in the early 2030s. Thinking about them now would be a good idea. If the programme to develop 100 gigawatts (GW) of nuclear power plants takes off and is on track in the 2030s, then closure of all thermal plants could be considered. This seems achievable and affordable. One advantage of this approach would be that the economy would have absorbed the costs of the closure of all thermal plants. It would therefore be in a better position to absorb the much higher costs of moving to net zero in the other hard-to-abate sectors of the modern industrial economy.

Achieving a carbon emission-free electricity system would be a major achievement and breakthrough. The journey to net zero thereafter would appear less challenging.

Transport

India's transport emissions are over 9% of the total emissions. Electrification would drive the reduction of emissions from the transport sector. This is the policy approach in Europe and California where they have decided that by 2035, sales of automobiles with internal combustion (IC) engines using fossil fuels would end. [4] All the auto majors in Europe are coming out with better and better electric versions of their popular models.

In the last few years, electric vehicles (EVs) are gaining visibility on Indian roads. This has been facilitated by government policies under its Faster Adoption and Manufacturing of Electric Vehicles (FAME) programme of providing financial support for the transition towards electric mobility (see Annexe 6.A.). India was forward looking in envisaging this transition and initiating work on it as early as the beginning of the last decade. The market share of EVs is rising. The Indian consumer is being offered an increasing number of EVs. Consumers are finding that their per kilometre cost of running an EV is so much lower than the petrol version. In the two-wheeler and three-wheeler segment, EVs have seen a rapid increase in market share over the last few years (Figure 4). In this segment, startups have taken the lead and dominate the market. This is an unexpected development and an illustration of the new vibrant startup culture that has emerged in India.

Japanese firms—Suzuki, Toyota, and Honda—have a major share of the Indian car market. The Japanese auto industry was late in seeing the emerging global market for EVs. They are trying to catch up. As soon as they come into the market with their EVs, the market share of EVs in India would rise even faster. The public charging infrastructure is steadily expanding and increasing the confidence of potential buyers of EVs.

Government took the lead in introducing electric buses by inviting bids for the running of 10,000 electric buses across the country with the bid parameter being the cost per kilometre.[5] The successful vendor had to put up the charging infrastructure, for which land and facilitation was assured, and run and maintain the buses. Aggregation of demand on this scale was expected to lead to lower prices. The results were a real surprise. The cost per kilometre turned out to be lower than the cost per kilometre of diesel buses, a real breakthrough. India is now moving rapidly towards greater use of electric buses. The Delhi Chief Minister has announced that all buses in Delhi would be electric by 2027. Such a decision could be extended to all new commercial vehicles across cities and regions having high levels of air pollution.[6] This would be politically easier as private vehicle owners would continue to have freedom of choice in the market and commercial vehicle operators would gain financially due to lower operating costs. The major immediate benefit would be the reduction of air pollution which has reached health crisis levels.

The rising share of EVs in India is represented in Figures 4 and 5.

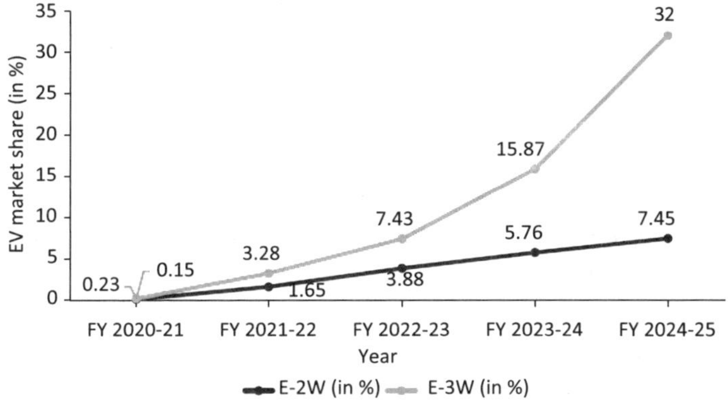

Figure 4 EV penetration trend for two- and three-wheelers[7]

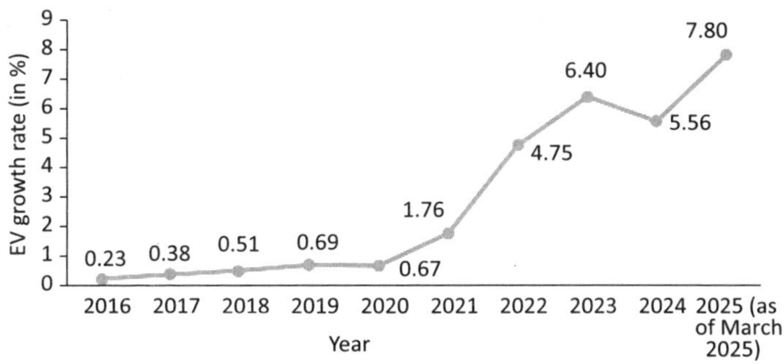

Figure 5 EV penetration by year in India [8,9]

Blending of biofuels is also being promoted in India for use by vehicles. This is, however, like energy efficiency, only an intermediate option to lower carbon intensity. The final objective is to have zero emissions. Biofuels are a renewable source of energy. Vehicles which run only on biofuels have net-zero emissions. Biowaste, cow dung, is a renewable source of energy which is available in abundance. It can be used for transport.[10] The technology for using this biogas for transport exists. But it would need a critical mass of a coordinated effort to create the full supply chain of tractors and other vehicles running on biogas, assured supply of biogas, and a sustainable business model. It would be better to focus on these rather than ethanol. A sustainable and scalable business model has to emerge from pilot projects. These need to be initiated. If price signals for biofuel lead to diversion of land-growing food to crops for biofuels, then this is an outcome which must be avoided. India just does not have the land and water endowments for growing crops for biofuel.

The milestones to lower emissions of road transport would be the following:

4. The share of EVs in new sales increase.
5. All new vehicles on the road become electric.
6. The share of renewables in electricity rises. Per passenger kilometre emissions begin to decline.
7. All heavy-duty trucks become green hydrogen trucks as electric heavy-duty trucks have not yet been developed.
8. Emissions from road vehicles peak as emissions from electricity peak.
9. Emissions decline, with the retirement of existing internal combustion-based vehicles. Decline would be faster as commercial vehicles which have the substantive share of emissions have a shorter useful life of about six years.
10. Emissions become zero when electricity emissions become zero.

To summarize, the peak and decline of carbon emissions to zero from road transport would happen when internal combustion engines using fossil fuels are no longer on the roads and carbon emissions from electricity generation end.

The number of vehicles would thereafter keep rising as the economy grows but with emissions remaining zero.

The Indian Railways have announced their intention to become net zero by 2030. This should not be difficult. They would be achieving this by:[11]

- Completion of full electrification of all passenger and freight movement.
- Only using electricity with zero emissions through captive solar power generation on their own land and buying carbon-free electricity for all their needs.

In shipping and civil aviation, which are global public goods, India would be moving along with other countries in reducing emissions and then taking them to zero. In shipping, green ammonia and hydrogen are technically feasible solutions for zero emissions. These are being tried out. Which of these two is the better solution would become clearer with time. Ships with engines using green fuels would have to be built to replace the existing global fleet.

India could choose to be smart and strategic and move into the building of green ships. Its human resource endowments give it the potential to succeed. Government should steer this by having a long-term programme of buying green ships made in India with increasing value addition. This kind of de-risking as suggested earlier for this full addition value chain of solar panels taking place in India could be adopted here. The potentially high cost of these ships could be passed on to the shipping costs by government mandates.

In civil aviation, the transition has to go beyond the ongoing efforts at producing and blending sustainable aviation fuel (SAF). Airbus has announced plans to develop aircraft using green hydrogen. Hopefully they would succeed, sooner than later. The transition to emission-free commercial travel would then begin.

Cooking

With a population of over 1.4 billion, the energy needs of cooking are large. About 6% of the total emissions come from cooking.[12]

The transition from the use of biomass to clean energy for cooking is nearing completion. Liquefied petroleum gas (LPG) is the main energy source for cooking. The use of piped natural gas (PNG) has been growing in urban areas as city gas distribution networks are being created and expanded.

To get to net zero, the use of gas for cooking would, however, have to end. To achieve this, replacing the use of gas by electricity for cooking would be necessary. This is relatively easy. Electricity is used in many parts of the world for cooking. Every home in India is now using electricity. It is at present used as a supplementary source for specific needs such as electric kettles for boiling water and microwave ovens for heating. The use of electric stoves in place of gas for all cooking has to take place.

Ten million homes are being provided with solar PV panels which would give the household about 300 units of free solar power every month. This is under a new programme started by Prime Minister Modi in 2024. It may be reasonable to expect the programme to be expanded later to cover all similarly placed households over time.[13] Some states are giving 300 units of free electricity per month to households. The lowest tariff slab for domestic consumers across states is well below the cost of supply. For all the households getting free electricity, the price signal for the use of electricity for cooking is strong.

But change in consumer perception and experience is a prerequisite for large-scale switch over from gas to electricity for cooking. The promotion of the use of electricity for cooking needs well-designed communication strategies. The confidence that Indian food can be made as well with electricity as with gas has to be created. Better electric cookstoves need to appear in the market. They should aim to increase market share with effective advertising of how their use would lead to savings. The sooner this programme is initiated, the better. Emissions from cooking could then become zero at the same time as electricity emissions become zero.

The use of biogas as a renewable source of energy for cooking also has potential. But production and reliable supply of biogas suitable for cooking along with a viable business model needs to be successfully demonstrated in some pilot projects. Scaling up would become a feasible option only after the successful implementation of pilot projects with consumer satisfaction.

In all these transitions in electricity generation and in the sectors to be electrified, the emissions would rise, peak, decline, and become zero.. The reduction from present level of emissions would be:

Category	CO_2 emission (in $MTCO_2e$) in 2020
Electricity production	1,166
Road transport	270
Railways	4
Total	**1,440**

Out of a total of 2,958 million tonnes of emissions in 2020, 1,440 million tonnes of emissions would become zero. About 51% of the emissions in 2020 would have been reduced. This would be a huge achievement.

But as the Indian economy would be growing in the coming decades, total emissions would also be rising. It may well be the case that even when the above sectoral emissions become zero. India's emissions would be considerably more than what they are at present. Getting to net zero in other sectors would however be a quite challenging. It is therefore essential that these sectors get attention now. This has, fortunately, begun to happen.

Industrial Decarbonization

As India makes its transition to becoming a developed economy, industrial production would keep rising. Emissions would therefore continue to grow though the increase would be moderated by the reduction in the carbon intensity of production. Emissions would peak only when all additional industrial production is done without carbon emissions. The emissions would remain at their peak till the time that the existing carbon-emitting plants using fossil fuels continue to produce. The emissions would begin to decline only when these start getting closed.

To shift to a greener path and decouple growth from increase in emissions would be a difficult challenge for each sector. The role of government policy would be decisive. Shifting to a greener, less carbon-intensive growth path is fortunately happening. Energy efficiency has been a low-hanging fruit. Many investments in energy efficiency are commercially viable with a short payback period. Once life cycle costs are compared instead of only the initial capital cost, most energy efficiency investments become profitable. The price signal of the short payback period has been very effective in the case of the highly energy-efficient LED light bulb replacing the traditional incandescent bulb.[14] Prices of LEDs were also brought down dramatically by repeated centralized bulk procurement for distribution companies (DISCOMs) who in turn gave these LED bulbs to their consumers. This approach of aggregating demand and undertaking bulk procurement by a government agency resulted in the increase in the scale of production in India and lower costs. Competition brought down costs further. This should be replicated wherever it can be made feasible.

Exceptional results in energy efficiency have been achieved through well-conceived pragmatic programmes evolved with stakeholder consultation. The energy intensity of India's GDP has declined by a remarkable 36% by 2020 from its level in 2005 (Annexe 6.B.).[15] India is now moving from success in energy efficiency to the next level of directly reducing carbon intensity. A carbon market is being put in place. The approach of the successful energy efficiency, Perform, Achieve, and Trade (PAT) scheme of the Bureau of Energy Efficiency (BEE) is being extended by it to the reduction of carbon intensity. Industry associations have been evolving strategies for this purpose (Annexe 6.C.).[16]

It needs to be recognized that carbon price is a mechanism to tweak market forces in the direction of the energy transition towards lower emissions. It is effective, but only at the margin. The experience in the advanced countries which adopted this policy instrument many years back has revealed its limitations. Can more be achieved? Yes.

The state, by taking greater responsibility, can accelerate change by creating additional, effective incentives and disincentives for market participants. For this it would have to be creative and innovative in using all the instruments at its disposal. These would include direct repetitive bulk procurement to get movement down the cost curve as in the case of the Solar Mission. Or, it could financially support a nascent technology in the market as is being done for EVs. Then it could impose specific enforceable regulations from a prospective date giving enough time for the markets to prepare and adjust.

R1 in Figure 6 is the carbon intensity reduction trajectory with ongoing measures and a carbon market. This trajectory

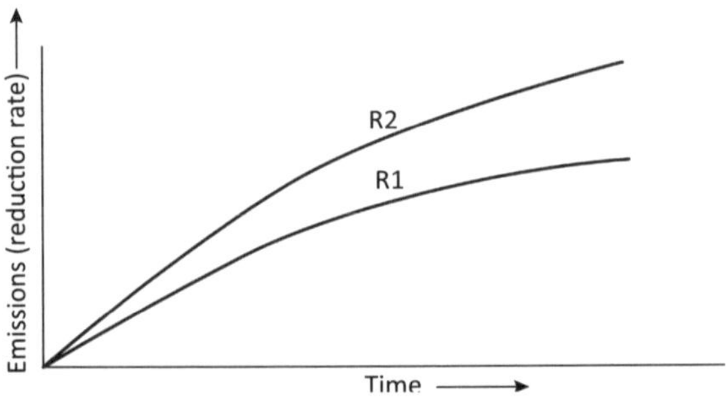

Figure 6 Carbon intensity reduction trajectory

would become much sharper, R2, with the state assuming greater responsibility and providing resources:

India's industrial emissions are about 628 $MTCO_2e$. This is about 21% of the total emissions. The following six large industrial segments account for the substantial share of industrial emissions.

1. Iron and steel
2. Cement
3. Aluminium
4. Ammonia and fertilizer
5. Pulp and paper
6. Petrochemicals

Industrial processes, especially in these, have greater potential for reducing or even eliminating carbon emissions than may appear likely at first sight. Industry-wise strategies evolved through consensus with stakeholder consultation, and which keep getting fine-tuned with the experience of implementation would make the transition smoother. This process has fortunately been initiated. Indian industry is accepting the need for this transition and trying to evolve a consensus on the feasible way forward.

Conceptually the use of fossil fuels and carbon emissions in industry can be reduced by a combination of the approaches discussed here.

Electrification

To the extent it is technically feasible, electricity should be used to replace fossil fuels in industrial processes. This substitution could be full or partial, depending on the technology in use. For example, the share of steel production through electric arc furnaces using scrap steel could be increased to the extent possible. New aluminium smelters could avoid putting up captive coal-based power plants. They could be asked to go in for buying electricity with reliable quality supply being assured. As electricity production moves towards becoming carbon free, these industrial processes would, in parallel, start becoming carbon free.

Solar thermal

Solar thermal concentrates the sun's rays to generate heat. Temperatures over 400 degrees Celsius can be generated. This heat can be used by industrial plants. These plants then need not use fossil fuels to generate heat for their manufacturing processes. This has enormous potential as a substitute for fossil fuels and is a low-hanging fruit. According to one study, nearly two-thirds of the energy consumption in Indian industries is utilized for generating heat to meet processing requirements with huge potential in industries such as food processing, textile dying, pharmaceuticals, and chemicals. But the use of solar thermal heat in industry is still at an early stage. It is cheaper to use this in new plants than in older plants by retrofitting, a good example of the late-mover's advantage. The possibilities are higher due to the availability of abundant sunlight in India, giving it a natural advantage over countries located in temperate zones. Full benefit of this natural endowment should be derived. This needs to be promoted with appropriate policies which should be evolved in partnership with industry. The setting up of a pilot plant for each segment through a special-purpose vehicle (SPV) would be a good beginning. The government should provide a subsidy to the extent needed to neutralize the cost difference between the market discovered price of solar thermal energy and the cost of the conventional fossil fuel heat energy for the industrial unit. The new industrial plant using solar thermal energy should not have to incur a higher cost. Depending on the cost differentials for different industrial processes, the shape of the optimal future pathways with cost implications would emerge. It is quite likely that lifetime costs would make solar thermal energy cost competitive or even cheaper as there are no fuel costs. Long-term heat supply contracts with repeated competitive bidding may lead to a substantial reduction in prices as has been seen with solar power. A new Solar Thermal Mission is called for. A dedicated mission should have the manpower and financial resources to be able to work with industry for this win-win transition.

For some sectors it may be worthwhile to impose binding regulations on the new plants that would be built. As suggested earlier, new aluminium plants should not be allowed to put up a captive coal-based thermal power plant, and they should be supplied reliable quality power supply for which they may be given a special dispensation of availing open access supply. Their emissions would decline as emissions from electricity generation declines. For steel the regulation could be that all new steel plants after a certain point in the future be built using gas-based direct-reduced iron (DRI) process using natural gas or syngas from coal gasification. The big advantage would be that with modest investment in retrofitting they could at some time in the future start using green hydrogen with 90% reduction in carbon emissions.[17] The cost for new plants would be a bit higher. But the Indian economy can bear this higher cost. There would be a level playing field for all new investors. If necessary, some import-duty protection could be provided to domestic producers.

Zero-emission Plants

With all the feasible successes through reduction in carbon intensity, if the underlying fossil fuel using technology continues being used for production, then emissions would keep rising. For the achievement of net zero, emissions from the use of fossil fuels must end altogether. This is more difficult and challenging. Recognizing this difficulty the phrase 'hard to abate' emerged. The creation of zero-carbon-emitting industrial plants and processes must remain the focus and get priority. It needs to be accepted that retrofitting an existing plant to become a zero-emission plant may not be technically feasible in many cases.

The challenges for the building of zero-emission plants are:

a. Higher cost: The products manufactured without emissions from the use of fossil fuels would usually be more expensive than the ones in the market using fossil fuels.

b. Many of the technologies for doing away with the use of fossil fuels need pilot projects to establish the technology for large-scale adoption. Costs must be discovered.

c. The potential for getting lower costs from larger volumes can then be estimated. Building the first pilot plants would need government support. The risk perception would usually be too high for the market to finance such plants. But government can afford to take the risk of time and cost overruns inherent in building a plant with a new technology that is yet to be used in a commercial-scale plant. Further, government can afford to provide the subsidy to meet the difference between the higher cost of production of a new zero-emission plant and the prevailing market price of existing plants using fossil fuels.

There would be many plants using fossil fuels. The transition to mandating that all new plants be built with zero-carbon emissions from a prospective date would pose a major challenge for policymakers. So much would depend on the costs at that time. Government would have to decide on how much of the higher cost should be passed on to the consumer, how much the government would absorb through lower tax rates, and how much through explicit subsidies. The extent to which the higher costs would be passed on to the consumer would be a political call. The extent to which the higher cost is absorbed by the government through a subsidy would depend on the fiscal space available at that time. Deciding on a policy for the closure of existing plants thereafter would also be challenging.

For the hard-to-abate sectors, it is too early to see the contours of the journey to net zero. Pilot projects for downstream use of green hydrogen in different segments need to be first implemented as suggested in the previous chapter. Some clarity on the higher cost resulting from giving up the use of fossil fuels as well as the potential for cost reduction with scale would emerge on completion of pilot projects. Affordable pathways to net zero in each of the hard-to-abate sectors could then be analysed under alternative scenarios. Evolving least-cost strategies would then be easier.

On the face of it these costs may appear to be too high today. The expectation that with competitive deployment and scale, costs would come down is reasonable. Further, depending on the costs in different sectors, government would be able to choose sectors to be taken up first as well as the pace of transition. There would be flexibility on the pace of change and cost absorption. Ideally, after the first few new zero-emission plants are set up, the mandate that all new plants to be built henceforth from a prospective date should be zero-emission plants would have to be considered. Peaking of emissions would take place from the time all new plants being built are zero-emission plants.

It needs to be noted that a mandate for putting up only carbon-free new plants in a sector, say sometime in the early 2030s, would put India at a theoretical advantage. As India is still at an early stage of industrialization, most of the plants in each sector have still to be built as the country moves towards becoming a developed nation. The sooner the year from which new plants would be carbon free, the lower would be the cost of closing carbon-emitting plants. The more industrialized the country the higher its cost in closing/retrofitting its existing large stock of fossil fuel using industrial plants in comparison to a country which is yet to build the bulk of its industrial plants. This is the essence of the "late-movers' advantage".

Residual Emissions

Industry

A few processes may not be amenable to change resulting in zero emissions. Their emissions would be marginal with reference to the total emissions which would be eliminated. These are industrial segments where hydrogen and electricity cannot be a full substitute. Cement is one such example. No substitute is available to replace cement, and no technological advancement has taken place to fully decarbonize its manufacturing process. As of now, with the use of alternative cementitious material, reduction of emissions by 30%[17] may be feasible. Carbon capture, for use or storage, is the only option for ending all carbon emissions in the cement industry. Carbon capture is achieved by separating carbon dioxide produced by the burning of a fossil fuel in an industrial plant. This technology exists. There is a cost to this capture of carbon dioxide. This carbon dioxide can be provided for industrial processes which need it as an input. This is usage. But the industrial-scale demand is very modest for carbon capture and use (CCU). The captured carbon can be injected inside the earth at suitable places into underground formations where the carbon dioxide can stay in perpetuity. This is expensive as the carbon dioxide must be piped and then injected inside the earth. India has been forward looking in recognizing this eventual need. The National Institution for Transforming India (NITI) Aayog, the government's think tank, has after expert consultations, prepared a strategy for this (Annexe 6.D.).

Agriculture

The agriculture sector contributes 13% of the total national GHG emissions which is about 384.8 $MTCO_2e$. Cattle and rice cultivation are the main sources of emissions.

According to India's fourth biennial update report to the UNFCCC, emissions from livestock accounted for about 222.63 $MTCO_2e$. The total livestock population is over 535 million and rising and so are the emissions. The mitigation options for reducing emissions are modest with changes in the feed. There is, however, no technical solution for complete mitigation of these emissions.

Emissions from rice cultivation are 67.73 $MTCO_2e$. There are mitigation techniques which could potentially lead to a reduction of 30%–50% in emissions. These also provide the benefits of reduction in water use by around 10%–20% as well as improving soil health without reducing yields. But getting farmers to change their traditional ways of cultivation is challenging and progress with sustained efforts may still be slow. These emissions can be reduced at best but not eliminated. Hence, sinks would be needed to achieve net zero for these (Annexe 6.E.).

Carbon Capture

Carbon capture and storage (CCS) and carbon capture utilization and storage (CCUS) are options which appear, prima facie, attractive. These may need to be utilized for the last steps in the journey to becoming net zero. But these technologies can only provide a solution, at the margin, for modest quantities of emissions. They just do not have the potential for providing substantial mitigation. The arithmetic of the total emissions and the potential for capture and use or capture and storage under the earth's surface just do not add up. The sooner this is accepted and assimilated in the discussion on getting to net zero, the better it would be. Thermal power generation cannot go on with carbon capture. Steel production has to move to using green hydrogen.

Conclusion

The pathways to net zero in the major carbon-emitting sectors are best envisaged through the prism of decoupling of growth and increase in energy consumption from further increase in carbon emissions. These would need to plateau. Then they must decline to become net zero. Alternative pathways with different timelines have to be delineated, with costs. The least-cost pathway in net present value terms becomes the rational choice. This exercise needs to be continuously updated as technology and costs keep evolving.

Peaking of emissions from electricity generation, and consequently in sectors, such as road transport, which get electrified, would result in real savings. This is such a positive beginning of the journey to net zero.

All other milestones on the journey for each sector have costs. Policies must aim to minimize these costs. At the same time, they need to aim to bring medium- to long-term benefits from the transformation into a green carbon-free sector. Being carbon free would generate a price premium in some segments of the domestic market and even more so in global markets. To illustrate, a tourism destination where all transport is green would attract more tourists. Manufactured exports would get a price advantage in Europe as their import duties linked to higher carbon emissions in production (CBAM) would not have to be paid.

If India moves swiftly into green manufacturing, it could achieve the breakthrough in export of manufactured goods that has eluded it so far. The costs of the journey to net zero could legitimately be seen as the costs of industrial policy to make India an advanced industrial economy. The capacity of the economy to absorb these costs if gradually imposed is generally underestimated. Given the easy transition on the electricity side, the gradual imposition of the higher costs in the industrial sectors should not impose shocks or have an adverse impact on economic growth. The key would be for the government to provide a predictable level playing field for all firms in a sector. On peaking of emissions in a sector when all additional production has zero emissions, there would be the need to have a CBAM type of import duty protection for domestic producers incurring a higher cost for green production. The benefits then would far outweigh the costs. There is a convergence in the paths to becoming net zero and an advanced green industrial economy. This convergence is real. India could reap the advantage of being a late mover into rapid industrialization and an early mover into green manufacturing. The Indian economy could take off with rapid green industrialization.

The journey to net zero would be long and complex. It appears daunting but it is achievable. Making the right policy choices is never easy. Making the correct choices and policies at key junctures starting from now would be the key. Achieving the goal of becoming net zero would then become both feasible and affordable. By 2070, certainly, or even earlier, quite likely. And what is most important is that on becoming net zero India would also become a globally competitive advanced green industrial economy. It would achieve its twin trysts with destiny of becoming a developed economy and doing all that it can to prevent the disaster unfolding from the global warming. India owes it to future generations to succeed in becoming net zero with focus, full commitment, and determination.

Annexes

Annexe 6.A.: FAME Scheme

The Faster Adoption and Manufacturing[18] of Electric Vehicles (FAME) programme for promotion of EVs. This scheme has evolved through three phases:

Phase I (2015–19)

- Phase I offered purchase subsidies for two-, three-, and four-wheelers, and buses.
- It supported pilot-charging infrastructure and R&D.

Phase II (2019–24)

- With a larger Rs 10,000 crore outlay, Phase II provided higher incentives for 2Ws, 3Ws, and e-buses, and funded widespread charging infrastructure.
- It encouraged local manufacturing.

PM E-DRIVE[19]

PM Electric Drive Revolution in Innovative Vehicle Enhancement (PM E-DRIVE) was notified in 2024 with an outlay of Rs 10,900 crore. Its aim is to support EVs including e-2W, e-3W, e-trucks, e-buses, e-ambulances, EV charging infrastructure, and the upgradation of testing agencies.

Annexe 6.B.: PAT Scheme for Promoting Energy Efficiency

Parliament was informed of the achievements of the PAT scheme in reply to a question.

The PAT Scheme was launched in 2012, and as of FY 2024, a total of 1,333 units across 13 energy-intensive sectors have been covered. This programme has saved 27.07 million tonnes of oil equivalent (MTOE) energy and encouraged investment of about Rs 75,000 crore in energy-efficient technologies and related measures.

The sector-wise details of energy target and savings are given below:

S. No.	Sector name	Energy saving target (in MTOE)	Achieved savings (in MTOE)	Monetary savings (in Rs crore)
1	Aluminium	1.057	2.138	4,806
2	Cement	2.183	3.474	7,006
3	Chlor and alkali	0.161	0.230	505
4	Fertilizer	0.924	1.163	1,998
5	Iron & steel	4.618	6.168	12,939
6	Pulp & paper	0.297	0.650	1,328
7	Textile	0.234	0.328	666
8	Power (thermal)	7.138	7.726	15,807
9	DISCOM	4.675	2.423	7,231
10	Refinery	2.267	2.261	5,150

Contd...

S. No.	Sector name	Energy saving target (in MTOE)	Achieved savings (in MTOE)	Monetary savings (in Rs crore)
11	Railways	0.077	0.196	585
12	Petrochemical	0.230	0.284	380
13	Building	0.009	0.031	42
	Grand total	23.87	27.07	58,444

The Bureau of Energy Efficiency (BEE) periodically conducts capacity enhancement programmes for designated consumers through various initiatives, including PAT-awareness programmes, sector-specific workshops on energy-efficient technologies, best practices, energy-efficient approaches, and process optimization. The BEE has organized 65 cross-sectoral workshops-cum awareness programmes and 9 sector-specific trainings on technology and best practices for the various stakeholders during 2024–25 (till February, 2025).

Ministry of Power has already issued notification in June 2023 for the addition of 13 new energy-intensive sectors under PAT Scheme including sectoral energy threshold level, namely sugar, chemical, ceramic, glass, zinc, copper, dairy, port trusts, automobile assembly units, tyre manufacturers, forging, foundry, and refractories.

The Baseline Audit of 280 units under newly notified sugar sector has been completed.

Further, the proforma development in the zinc and copper sector has been completed.

Annexe 6.C.: Carbon Credit Trading Scheme

The Carbon Credit Trading Scheme (CCTS) builds on the earlier PAT Scheme to reduce industrial emissions. The Scheme uses a 'baseline-and-credit' approach:

Each industrial facility is assigned a baseline GHG emission intensity for 2023–24.

Annual targets for reducing emissions intensity (CO_2 per unit of production) are set.

Lower-than-target emissions earn carbon credits.

Higher-than-target emissions require purchase of credits or payment of a penalty.

CCTS coverage and targets (2025–27)

Sector	No. of Plants	Average reduction (2025–27)	Sector-specific observations
Cement	186	3.4%	Contributes ~6% of industrial emissions. Receives the mildest targets despite large share.
Aluminium	13	5.85%	Accounts for ~2% of emissions. Moderate targets, slightly above previous PAT benchmarks.

Contd...

Sector	No. of Plants	Average reduction (2025–27)	Sector-specific observations
Pulp & paper	53	7.15%	Each contributes <1% to emissions. Assigned the most aggressive cuts, likely because emission reductions here are more feasible and cost-effective.
Chlor alkali	30	7.54%	

Annexe 6.D.: Carbon Sinks

A carbon sink is any system—natural or engineered—that absorbs more carbon dioxide (CO_2) from the atmosphere than it releases. Forests, soils, mangroves, agroforestry, wetlands, and even ocean ecosystems are the world's largest natural sinks.

Increasing the nation's carbon sink is essential for the achievement of net zero. Sinks are needed to offset dispersed carbon emissions which cannot be eliminated. For instance, emissions from rice cultivation can at best be reduced but not eliminated. Under its Nationally Determined Contributions (NDCs) India's commitment under the Paris Agreement is for the creation of an additional carbon sink of 2.5–3 billion tonnes of CO_2e by 2030.

India has one of the highest population densities in the world with an adverse land–man ratio. With great difficulty it was able to reverse the earlier trend of declining forest cover with strong commitment, policies, and programmes. But increasing the forest cover substantially does not appear feasible.

Sinks can be increased by:

a. Planting more trees (afforestation/reforestation)
b. Protecting and expanding wetlands, mangroves, grasslands
c. Improving farm practices (agroforestry, biochar)

There are ongoing programmes for each of these.

Agroforestry has been making good progress as the private sector and the farmers are being able to join hands in mutually beneficial and commercial arrangements.

A policy change under which the private sector is used for the densification of the tree cover in designated forest areas with appropriate species has the potential of delivering better outcomes.

But the overall potential is modest. Further, ground realities make progress challenging and difficult.

Annexe 6.E.: Carbon Capture, Utilization, and Storage (CCUS)

India has recognized the necessity of carbon capture and use, and storage for absorbing the emissions from the hard-to-abate sectors where as of now elimination altogether does not appear feasible. The NITI Aayog has done pioneering work and prepared a holistic strategy.

Estimated CO_2 storage potential in India

Geological formation	Estimated potential
Onshore sedimentary basins	395 $GTCO_2$
Offshore formations	100 $GTCO_2$
Depleted oil and gas reservoirs	~5–10 $GTCO_2$
Coal seams (ECBM)	~50 $GTCO_2$ (est.)
Basalt formations	~100–300 $GTCO_2$

The following pilot plants for technology demonstration and cost discovery are under development:

Company / consortium	Sector and technology	Scale/TRL	Location and timeline
NTPC + Jakson Green + Carbon Clean	Post-combustion amine capture → catalytic methanol synthesis	20 T CO_2 d^{-1} pilot; TRL 7	Vindhyachal (Madhya Pradesh) – Phase-1 commissioned Nov. 2024
Dalmia Cement + SaltX	Electric arc calciner + oxy-fuel capture	100 t clinker d^{-1} demo; TRL 6–7	Tamil Nadu pilot by 2025

Contd...

Company / consortium	Sector and technology	Scale/TRL	Location and timeline
Oil India + UGreen Tech (start-up)	Modular direct flue-gas CO_2 capture skids	3 T CO_2 d^{-1}; TRL 6	North-east fields – online 2025
Public Testbed Cluster (5 sites)	Catalysis, advanced solvents, vacuum swing, algae utilization	Mixed TRL 4–7	Funded Jan. 2025 under PPP CCU Testbeds scheme

Technology type	Sector/ application	TRL in India	Cost range (in USD/ TCO_2)	Lead stakeholders
Solvent-based capture	Coal power, cement	7–9	60–90	NTPC, BHEL, CSIR
Pressure swing adsorption	Refineries, ammonia	6–7	70–100	IOCL, RCF
Calcium looping	Cement	5–6	80–110	Dalmia Bharat, IITs
CO_2 mineralization	Aggregates, cement	7	<50	Dalmia Cement, startups
Bio-CCUS (algae, biofuels)	Bioenergy, agriculture	5	Not commercial	CSIR, ICAR

Utilization

Utilization pathway	End-products	TRL	Industrial partners
CO_2 + Green H_2	Methanol, urea, DME	6–8	RCF, GSFC, NTPC, Jakson
Mineralization	Construction materials	7	Dalmia Cement, CarbonCure
Algae-based CO_2 fixation	Biofuels, bioplastics	4–5	CSIR-NEERI, private startups

Dalmia Cement is developing carbonated cement and aggregates, with internal target of 5 $MTCO_2$ reduction by 2030.

Institution	Flagship project	Year/ TRL	Output
IIT Bombay + DST-SERB	Super-base ionic liquids for low-grade heat capture	2025, TRL 5	35% solvent energy cut
JNCASR Bengaluru	Nano–MgO looping for cement flues	2024, TRL 4	12% △ capture-efficiency
CSIR-NEERI	Bio-mineralization using cyanobacteria	2025, TRL 4–5	1 kg CO_2 → 0.6 kg $CaCO_3$
IISc	Basalt core carbonation kinetics (Deccan)	2025, TRL 6 (pilot core)	80% mineral trapping in 2 years

Costs

Total costs typically range between \$75 and 110/TCO_2 for end-to-end systems.

Segment	Cost (in USD/TCO_2)
Capture (solvent)	50–70
Compression and dehydration	10–15
Transport (pipeline)	10–15
Storage/injection	5–10
Total	75–110

References

Chapter 1

1. Sarma, N. A. 1958. Economic development in India: the first and the second five year plans. *IMF eLibrary*. DOI: https://doi.org/10.5089/9781451949636.024.A002
2. Narasimham, M. 1962. India's Third Five-year Plan. IMF eLibrary. DOI: https://doi.org/10.5089/9781451968873.024.A005
3. *Ministry of Statistics and Programme Implementation* (MoSPI). 2018. *Five-year plans - Chapter 7*. Details available at <https://www.mospi.gov.in/sites/default/files/Statistical_year_book_india_chapters/ch7.pdf>
4. Government of India. 2003. The Electricity Act, 2003 (with amendments). Central Electricity Regulatory Commission. Details available at <https://cercind.gov.in/Act-with-amendment.pdf>
5. Ministry of Power. 2005. National Electricity Policy. Government of India. Details available at <https://powermin.gov.in/en/content/national-electricity-policy>
6. Burlig, F. and Preonas, L. 2021. Out of the darkness and into the light? Development effects of rural electrification. *Energy Institute at Haas*. Details available at <https://haas.berkeley.edu/wp-content/uploads/WP268.pdf>
7. Oda, H. and Tsujita, Y. 2011. The determinants of rural electrification: the case of Bihar, India. *Energy Policy,* **39** (6): 3086–3095. DOI: https://doi.org/10.1016/j.enpol.2011.02.014
8. Rehman, I. H. 2004. Rural energy in India. In *Elsevier eBooks,* pp. 507–514. DOI: https://doi.org/10.1016/b0-12-176480-x/00452-6
9. Office of the Registrar General and Census Commissioner, India. (2001). A-01: Number of villages, towns, households, population and area - 2001 [Data set]. Government of India. Details available at <https://censusindia.gov.in/nada/index.php/catalog/20028>

10. Lok Sabha Secretariat. 2000 (18 December). Lok Sabha debates: 13th Lok Sabha, 5th session. Parliament Digital Library. Details available at <https://eparlib.nic.in/bitstream/123456789/759545/1/lsd_13_05_18-12-2000.pdf>
11. Office of the Registrar General and Census Commissioner, India. (2001). H-09: Distribution of households by source of lighting (all). Census of India 2001. Details available at <https://censusindia.gov.in/nada/index.php/catalog/18628>
12. The World Bank. 2000. Implementation Completion Report: NTPC Power Generation Project. Details available at <https://documents1.worldbank.org/curated/en/164761468041411677/pdf/multi-page.pdf>
13. NTPC Limited. 2025. NTPC Group touches 400 billion units power generation mark in FY 2024–25. Details available at <https://ntpc.co.in/media/press-releases/ntpc-group-touches-400-billion-units-power-generation-mark-fy-2024-25>
14. NTPC Limited. *n.d. NTPC overview*. Details available at <https://ntpc.co.in/ntpc-overview>
15. Government of India. 2005. (25 November). Electrification of villages: starred Question No. 43 answered in Lok Sabha. Parliament Digital Library. Details available at <https://eparlib.nic.in/bitstream/123456789/533628/1/33933.pdf>
16. National Democratic Alliance. 2004. Agenda for development, good governance and peace: Elections to the 14th Lok Sabha. Bharatiya Janata Party. Details available at <https://library.bjp.org/jspui/bitstream/123456789/245/1/NDA%20MANIFESTO%202004.pdf>
17. Indian National Congress. 2004. Congress Manifesto 2004: Elections to the 14th Lok Sabha. Details available at <https://archive.org/details/CongressManifesto2004>
18. Shankar, A. 2025 (14 January). Nothing 'Weak' or 'Accidental' in the Decisiveness With Which Manmohan Singh Led India. The Wire. Details available at <https://m.thewire.in/article/politics/nothing-weak-or-accidental-in-the-decisiveness-with-which-manmohan-singh-led-india>
19. Standing Committee on Energy. 2006 (February). Tenth Report: Action taken on the recommendations contained in

the Fifth Report (14th Lok Sabha) on Demands for Grants of the Ministry of Power for the year 2005–2006. Lok Sabha Secretariat. Details available at <https://eparlib.nic.in/bitstream/123456789/62741/1/14_Energy_10.pdf>

20. Ernst & Young 2007 (5 March). Models of rural electrification: final report to the Forum of Indian Regulators. Forum of Indian Regulators. Details available at <https://foir-india.org/Final_Foir_Report_5th_March07_Models_of_RE_E%26Y%5B1%5D.pdf>
21. Power Line Magazine. 2016 (28 March). A long way to go: Reality check on rural electrification. Details available at <https://powerline.net.in/2016/03/28/a-long-way-to-go/>
22. International Energy Agency. 2017. *Energy Access Outlook 2017: From Poverty to Prosperity* (World Energy Outlook Special Report). OECD/IEA. Details available at <https://iea.blob.core.windows.net/assets/9a67c2fc-b605-4994-8eb5-29a0ac219499/WEO2017SpecialReport_EnergyAccessOutlook.pdf>
23. Ministry of Power. *n.d.* Saubhagya – Pradhan Mantri Sahaj Bijli Har Ghar Yojana. Government of India. Details available at <https://powermin.gov.in/en/content/saubhagya>
24. International Energy Agency. 2020. India 2020: Energy Policy Review. NITI Aayog. Details available at <https://www.niti.gov.in/sites/default/files/2020-01/IEA-India-In-depth-review2020.pdf>
25. Banerjee, S. G., Barnes, D., Singh, B., Mayer, K., and Samad, H. (2015). *Power for All: electricity access challenge in India.* World Bank. Details available at <https://documents1.worldbank.org/curated/en/562191468041399641/pdf/922230PUB0978100Box385358B00PUBLIC0.pdf>
26. Rohde, D. 2002 (16 December). India moves to overcome crippling electricity shortages. *The New York Times*. Details available at <https://www.nytimes.com/2002/12/16/news/international-business-india-moves-to-overcome-crippling-electricity-shortages.html>
27. Bolton, P. and Stewart, I. 2024 (10 December). Domestic Energy Prices (Commons Library Research Briefing No. 9491). House of Commons Library. Details available at <https://commonslibrary.parliament.uk/research-briefings/cbp-9491/>

28. Ajayi, V., Burlinson, A., Giulietti, M., and Waterson, M. 2024 (24 October). The impact of the energy price crisis on UK customers. Centre for Competitive Advantage in the Global Economy (CAGE), University of Warwick. Details available at <https://warwick.ac.uk/fac/soc/economics/research/centres/cage/news/24-10-24-the_impact_of_the_energy_price_crisis_on_uk_customers/>
29. Ministry of Power. *n.d.* Ultra Mega Power Projects. Government of India. Details available at <https://powermin.gov.in/sites/default/files/uploads/ultra_mega_power_project.pdf>
30. Mukherjee, M. 2014. *Private participation in the Indian power sector: lessons from two decades of experience.* World Bank. DOI: https://doi.org/10.1596/978-1-4648-0339-0
31. Skill Council for Green Jobs. 2016 (May). Clean cooking: identification of green business potential in India. Government of India. Details available at <https://sscgj.in/wp-content/uploads/2016/05/Concept_note-Clean-Cooking-Solutions.pdf>
32. Ministry of Petroleum and Natural Gas (MoPNG). 2025 (3 April). Government steps to ensure affordable LPG. Press Information Bureau, Government of India. Details available at <https://pib.gov.in/PressReleaseIframePage.aspx?PRID=2118208>
33. Ministry of Petroleum and Natural Gas (MoPNG). 2023 (20 April). Pradhan Mantri Ujjwala Yojana fuels LPG revolution: Nearly 17 crore LPG consumers added in last 9 years. Press Information Bureau, Government of India. Details available at <https://pib.gov.in/PressReleasePage.aspx?PRID=1918364>
34. Sanghera, T. 2019 (25 April). PM's Ujjwala LPG scheme will meet objectives if poor pay less for refills than middle class, rich. Scroll.in. Details available at <https://scroll.in/article/920900/pms-ujjwala-lpg-scheme-will-meet-objectives-if-poor-pay-less-for-refills-than-middle-class-rich>
35. Palit, D., Shardul, M., and Brahmachari, D. 2021. *Clean Fuel for Cooking: solution to achieve better air quality* (policy brief). The Energy and Resources Institute. Details available at <https://www.teriin.org/sites/default/files/2021-02/clean-fuel-for-cooking-solutions-pb_0.pdf>

36. Chadha, S. 2022 (10 January). Assembly elections: parties are promising 'free' power. Who will bear this cost? *The Times of India*. Details available at <https://timesofindia.indiatimes.com/business/india-business/explained-punjab-and-up-govt-are-doling-out-free-electricity-ahead-of-elections-but-who-will-bear-this-cost/articleshow/88804306.cms>
37. Press Information Bureau. 2025 (6 January). UJALA: 10 years of energy-efficient lighting. Government of India. Details available at <https://pib.gov.in/PressReleasePage.aspx?PRID=2090639>
38. Chadha, S. 2023 (12 October). Congress promises cow dung purchase scheme in MP if voted to power. *Telangana Today*. Details available at <https://telanganatoday.com/congress-promises-cow-dung-purchase-scheme-in-mp-if-voted-to-power>
39. Shankar, A. 2024 (4 January). Energy policies for India. India Foundation. Details available at <https://indiafoundation.in/articles-and-commentaries/energy-policies-for-india/>
40. Press Information Bureau. 2025 (13 March). PM Surya Ghar: Muft Bijli Yojana crosses milestone of 10 lakh installations. Ministry of New and Renewable Energy, Government of India. Details available at <https://pib.gov.in/PressReleaseIframePage.aspx?PRID=2111106>
41. Shankar, A. and Avni, T. C. A. 2021. *Resolving the Crisis in Power Distribution in India* (discussion paper). The Energy and Resources Institute. Details available at <https://www.teriin.org/sites/default/files/2021-08/power-distribution-India-dp.pdf>
42. Ministry of Power. *n.d. Saubhagya: Pradhan Mantri Sahaj Bijli Har Ghar Yojana*. Government of India . Details available at <https://powermin.gov.in/en/content/saubhagya>

Chapter 2

1. Sethna, H. N. 1979. India's atomic energy programme: past and future. *IAEA Bulletin* **21** (5): 9–11. International Atomic Energy Agency. Details available at <https://www.iaea.org/sites/default/files/publications/magazines/bulletin/bull21-5/21505090211.pdf>
2. Ministry of New and Renewable Energy, Government of India. *n.d.* Introduction. Details available at <https://mnre.gov.in/en/about-department/introduction/>

3. Ministry of New and Renewable Energy 2025 (10 April). India's renewable energy capacity achieves historic growth in FY 2024-25. Press Information Bureau. Details available at <https://www.pib.gov.in/PressReleasePage.aspx?PRID=2120729>
4. Ministry of New and Renewable Energy, Government of India. 2024. *Renewable Energy Statistics 2023–24*. Details available at <https://cdnbbsr.s3waas.gov.in/s3716e1b8c6cd17b771da77391355749f3/uploads/2024/10/20241029512325464.pdf>
5. Maguire, G. 2024 (22 August). Germany's top clean energy source set to slow growth pace in 2024. Reuters. Details available at <https://www.reuters.com/business/energy/germanys-top-clean-energy-source-set-slow-growth-pace-2024-maguire-2024-08-22/>
6. Sud, T., Sharma, R., Sharma, R., and Kitson, L. 2015. *India's Accelerated Depreciation Policy for Wind Energy* (case study). International Institute for Sustainable Development. Details available at <https://www.iisd.org/system/files/publications/india-accelerated-depreciation-policy-wind-energy-case-study.pdf>
7. Tamil Nadu Electricity Regulatory Commission. 2021 (5 August). Order in MP No. 21 of 2021. Details available at <https://www.tnerc.tn.gov.in/Orders/files/CO-MPNo21%20050820211743.pdf>
8. Gopal, S. 2023 (8 March). Competitive bidding raises hopes for wind sector. Centre for Science and Environment. Details available at <https://www.cseindia.org/competitive-bidding-raises-hopes-for-wind-sector-11663>
9. Ministry of New and Renewable Energy, Government of India. *n.d.* Wind overview. Details available at <https://mnre.gov.in/en/wind-overview/>
10. Climate Adaptation Platform. 2021 (2 June). Wind speeds in India are affecting wind energy generation. Details available at <https://climateadaptationplatform.com/wind-speeds-in-india-are-affecting-wind-energy-generation/>
11. Ministry of Law and Justice. 2003. The Electricity Act, 2003 (No. 36 of 2003). Government of India. Details available at <https://cercind.gov.in/Act-with-amendment.pdf>
12. Central Electricity Regulatory Commission. 2010 (14 January). CERC Regulation on Renewable Energy Certificates (REC).

Government of India. Details available at <https://www.cercind.gov.in/Regulations/CERC_Regulation_on_Renewable_Energy_Certificates_REC.pdf>

13. Yashodha, Y., Sanjay, A., and Mukherji, A. 2021. *Solar Irrigation in India: a situation analysis report*. International Water Management Institute. Details available at <https://solar.iwmi.org/wp-content/uploads/sites/43/2021/09/INDIA-SITUATION-ANALYSIS-REPORT_final-version-3.pdf>
14. Press Information Bureau. 2003 (23 July). PIB Press Release. *Renewable Energy Vision*. Details available at <https://archive.pib.gov.in/release02/lyr2003/rjul2003/23072003/r2307200324.html>
15. Pulipaka, S. 2019 (8 February). Solar in India: the 10-year challenge. *pv magazine India*. Details available at <https://www.pv-magazine-india.com/2019/02/08/solar-in-india-the-10-year-challenge/>
16. Ministry of New and Renewable Energy. (2010). Jawaharlal Nehru National Solar Mission: Towards Building Solar India. Solar Energy Corporation of India. Government of India Details. Details available at <https://www.seci.co.in/upload/static/files/mission_document_JNNSM(1).pdf>
17. Mercom India. 2018 (5 February). Mercom exclusive: solar tariffs in India have fallen by 73% since 2010. Details available at <https://www.mercomindia.com/mercom-exclusive-solar-tariffs-india-fallen-73-percent-since-2010>
18. Koundal, A. 2020 (30 June). India achieves historic low solar tariff of Rs 2.36 per kWh for 2-GW ISTS tender. *ET EnergyWorld*. Details available at <https://energy.economictimes.indiatimes.com/news/renewable/india-achieves-historic-low-solar-tariff-of-rs-2-36-per-kwh-for-2-gw-ists-tender/76714472>
19. Central Electricity Regulatory Commission. (2024). Annexure–I: Pro-forma for furnishing actual annual performance/operational data for coal/lignite-based thermal generating stations (2017–2022): Singrauli Super Thermal Power Station. Details available at <https://cercind.gov.in/2024/draft_reg/O&M_data/NTPC/Singrauli%20Final.pdf>
20. Press Information Bureau. 2013 (22 April). Viability gap funding for solar projects. Ministry of New and Renewable Energy, Government of India. Details available at <https://www.pib.gov.in/newsite/PrintRelease.aspx?relid=94876>

21. Press Information Bureau. 2017 (28 December). Schemes launched by the Government to promote Solar Energy in the country. Ministry of New and Renewable Energy, Government of India. Details available at <https://www.pib.gov.in/PressReleasePage.aspx?PRID=1514462>
22. Press Information Bureau. 2023 (28 September). India is committed to achieve the net-zero emissions target by 2070 as announced by PM Modi, says Dr Jitendra Singh. Ministry of Science and Technology, Government of India. Details available at <https://www.pib.gov.in/PressReleaseIframePage.aspx?PRID=1961797>
23. India Climate & Energy Dashboard. (n.d.). Installed capacity mix. NITI Aayog. Retrieved September 11, 2025, from https://iced.niti.gov.in/energy/electricity/generation/capacity
24. IndiaStat. (n.d.). Selected state-wise wind power installed capacity in India (1992–1993 to 2007–2008). Retrieved September 11, 2025, from https://www.indiastat.com/table/power/selected-state-wise-wind-power-installed-capacity-/12735
25. Press Information Bureau. (2023, August 9). Wind power potential of windy states and policy for repowering wind power projects [Press release]. Ministry of New and Renewable Energy. https://www.pib.gov.in/PressReleasePage.aspx?PRID=1947133
26. Ministry of New and Renewable Energy. (2025, September 11). Wind Overview. Government of India. https://mnre.gov.in/en/wind-overview/
27. India Climate & Energy Dashboard. (n.d.). Power generation mix. NITI Aayog. Retrieved September 11, 2025, from https://iced.niti.gov.in/energy/electricity/generation/power-generation

Chapter 3

1. Press Trust of India. 2025 (8 April). India overtakes Germany to become 3rd-largest generator of wind, solar power. Deccan Herald. Details available at <https://www.deccanherald.com/india/india-overtakes-germany-to-become-3rd-largest-generator-of-wind-solar-power-report-3483098>
2. International Energy Agency. 2025. Germany—countries and regions. Details available at <https://www.iea.org/countries/germany>

3. Fraunhofer Institute for Solar Energy Systems ISE. 2025 (15 January). Public electricity generation 2024: Renewable energies cover more than 60 per cent of German electricity consumption for the first time. Details available at <https://www.ise.fraunhofer.de/en/press-media/press-releases/2025/public-electricity-generation-2024-renewable-energies-cover-more-than-60-percent-of-german-electricity-consumption-for-the-first-time.html>
4. Central Electricity Authority. 2023 (May). National Electricity Plan (2022–2032). Ministry of Power, Government of India. Details available at <https://cea.nic.in/wp-content/uploads/irp/2023/05/NEP_2022_32_FINAL_GAZETTE-1.pdf>
5. Ministry of Power, Government of India. 2022 (22 July). Renewable purchase obligation (RPO) and energy storage obligation (ESO) trajectory till 2029–30. Details available at <https://powermin.gov.in/sites/default/files/Renewable_Purchase_Obligation_and_Energy_Storage_Obligation_Trajectory_till_2029_30.pdf>
6. Shankar, A., Saxena, A. K., Mazumdar, R. 2023 (November). *Pumped Storage Plants: essential for India's energy transition.* The Energy and Resources Institute (TERI). Details available at <https://www.teriin.org/sites/default/files/2024-01/Pumped_Storage%20_Plants_Discussion_Paper_2023.pdf>
7. International Hydropower Association. 2018. The world's water battery: pumped storage hydropower and the clean energy transition. Details available at <https://cdn.prod.website-files.com/64f9d0036cb97160cc26feba/64f9d0036cb97160cc27 0fc9_the_worlds_water_battery_-_pumped_storage_and_the_clean_energy_transition_2_1.pdf>
8. Power Line Magazine. 2023 (4 April). Hits and misses: expert opinion on the guidelines for promoting PSPs. Details available at <https://powerline.net.in/2023/04/04/hits-and-misses-expert-opinion-on-the-guidelines-for-promoting-psps/>
9. Shankar, A., Saxena, A. K., and Mazumdar, R. 2024 (February). *Concentrating Solar Power Plants with Storage: deployment essential now.* The Energy and Resources Institute (TERI). Details available at <https://teriin.org/sites/default/files/2024-02/CSP_Report_2024_New.pdf>

10. The World Bank. 2023 (January). Guidelines to implement battery energy storage systems under publicprivate partnership structures. Details available at <https://documents1.worldbank.org/curated/en/099536501202316060/pdf/IDU0edcfc32c0825f040f509c0b0bbf49294e569.pdf>
11. Ambri. 2023 (8 August). Ambri achieves UL 1973 certification for its next generation Liquid Metal™ battery cells. Business Wire. Details available at <https://www.businesswire.com/news/home/20230808943870/en/Ambri-Achieves-UL-1973-Certification-for-Its-Next-Generation-Liquid-Metal%E2%84%A2-Battery-Cells>
12. International Hydropower Association. 2023. *2023 World Hydropower Outlook*. Details available at <https://indd.adobe.com/view/4201016f-a51a-4f6f-998b-ec85219d1dfd>
13. Jacob, A. S., and Garg, R. 2021 (6 July). How India can accelerate pumped hydro storage for a clean energy future. Center for Study of Science, Technology and Policy (CSTEP). Details available at <https://cstep.in/drupal/sites/default/files/2021-07/OP-ED%20Pumped%20Hydro%20Ammu%20and%20Rishu.pdf >
14. Lyu, X., Zhang, T., Yuan, L., Yang, K., Fang, J., Li, S., and Liu, S. 2022. Pumped storage hydropower in abandoned mine shafts: key concerns and research directions. *Sustainability*, **14** (23): 16012. DOI: https://doi.org/10.3390/su142316012
15. Proctor, D. 2022 (5 January). Former coal mine will house new pumped-hydro storage project. *POWER Magazine.* Details available at <https://www.powermag.com/former-coal-mine-will-house-new-pumped-hydro-storage-project/>
16. Central Electricity Authority. 2023 (May). Status of pumped storage development in the country. Ministry of Power, Government of India. Details available at <https://cea.nic.in/wp-content/uploads/hpi/2023/05/Status_of_Pumped_storage_development_in_country-2.pdf>
17. Ministry of Power. 2024 (May). Sanction Order: Scheme for Viability Gap Funding for development of Battery Energy Storage Systems-reg. Details available at <https://cdnbbsr.s3waas.gov.in/s3716e1b8c6cd17b771da77391355749f3/uploads/2024/05/20240503164033573.pdf>

Further reading

1. Forum of Regulators. 2023 (August). *Hydro PSP Report*. Details available at <https://forumofregulators.gov.in/Data/study/Hydro%20PSP%20Report.pdf<
2. Ministry of Power, Government of India. 2024 (February). Draft guidelines for tariff-based competitive bidding process for procurement of power from pumped storage projects. Details available at <https://powermin.gov.in/sites/default/files/webform/notices/Draft_TBCB_PSP_Guidelines.pdf>
3. Ministry of Power, Government of India. 2023 (April). Guidelines to promote development of pumped storage projects. Details available at <https://powermin.gov.in/sites/default/files/webform/notices/Guidelines_to_Promote_Development_of_Pump_Storage_Projects.pdf>
4. Shankar, A. 2024 (18 February). Competitive bidding for pump storage plants development. *The Financial Express*. Details available at <https://www.financialexpress.com/business/sustainability-competitive-bidding-for-pump-storage-plants-development-3397221/>
5. The Energy and Resources Institute (TERI). 2024 (9 February). Concentrating solar power (CSP) plants with storage. Details available at <https://teriin.org/sites/default/files/2024-02/CSP_Report_2024_New.pdf>
6. The Energy and Resources Institute (TERI). 2023 (20 November). Pumped storage plants (PSP) – Essential for India's energy transition. Details available at <https://www.teriin.org/sites/default/files/2024-01/Pumped_Storage%20_Plants_Discussion_Paper_2023.pdf>
7. Shankar, A. 2023 (5 September). Energy transition: why it is time for India to focus on CSP with storage. *The Energy World, Economic Times*. Details available at <https://energy.economictimes.indiatimes.com/news/renewable/energy-transition-why-it-is-time-for-india-to-focus-on-csp-with-storage/103384575>
8. Shankar, A. 2023 (6 July). The bright side to power storage options. *The Hindu BusinessLine*. Details available at <https://www.

thehindubusinessline.com/opinion/the-bright-side-to-power-storage-options/article67050139.ece>

9. Shankar, A. 2023 (7 February). Pump storage plants: the way ahead for energy storage in India. *The Energy World, Economic Times*. Details available at <https://energy.economictimes.indiatimes.com/news/power/pump-storage-plants-the-way-ahead-for-energy-storage-in-india/97661561>
10. Shankar, A. 2022 (24 May). Future energy needs storage push today. *The Financial Express*. Details available at <https://epaper.financialexpress.com/3484071/Delhi/May-24-2022#clip/68210378/c99cd51a-2da9-4dc5-a4d7-665ba92ade83/810.6666666666665:1026.263102725367>

Chapter 4

1. Press Information Bureau. 2022 (3 February). India's stand at COP26. Ministry of Environment, Forest and Climate Change, Government of India. Details available at <https://www.pib.gov.in/PressReleasePage.aspx?PRID=1795071>
2. Murtaugh, D. 2021 (1 November). World's cheapest renewables will aid India's push to net zero. Bloomberg. Details available at <https://www.bloomberg.com/news/articles/2021-11-02/world-s-cheapest-renewables-will-aid-india-s-push-to-net-zero>
3. International Energy Agency. 2024 (9 October). Massive global growth of renewables to 2030 is set to match entire power capacity of major economies today, moving world closer to tripling goal. Details available at <https://www.iea.org/news/massive-global-growth-of-renewables-to-2030-is-set-to-match-entire-power-capacity-of-major-economies-today-moving-world-closer-to-tripling-goal>
4. Press Information Bureau. 2023 (25 August). Need to shift climate change narrative from total emissions to per capita emissions of each country: Union Minister for Power and New & Renewable Energy. Ministry of Power, Government of India. Details available at <https://www.pib.gov.in/PressReleasePage.aspx?PRID=1952260>

5. United Nations Framework Convention on Climate Change. *n.d. Article 3: Principles*. Details available at <https://unfccc.int/resource/ccsites/zimbab/conven/text/art03.htm>
6. Press Information Bureau. 2024 (2 December). Initiatives to harness hydro potential. Ministry of Power, Government of India. Details available at <https://www.pib.gov.in/PressReleasePage.aspx?PRID=2079832>
7. Central Electricity Authority. 2023 (May). Status of pumped storage development in the country. Ministry of Power, Government of India Details available at <https://cea.nic.in/wp-content/uploads/hpi/2023/05/Status_of_Pumped_storage_development_in_country-2.pdf>
8. FE Bureau. 2021 (2 July). India to add 26 GW hydro projects by 2030: IEA. *The Financial Express*. Details available at <https://www.financialexpress.com/business/industry-india-to-add-26-gw-hydro-projects-by-2030-iea-2282279/>
9. Ministry of New and Renewable Energy. 2025 (15 July). Physical achievements: programme/scheme-wise cumulative physical progress as on 30 June 2025. Government of India. Details available at <https://mnre.gov.in/en/physical-progress/>
10. Press Information Bureau. 2025 (10 April). India's renewable energy capacity achieves historic growth in FY 2024–25. Ministry of New and Renewable Energy, Government of India. Details available at <https://www.pib.gov.in/PressReleasePage.aspx?PRID=2120729>
11. Press Information Bureau. 2025 (13 March). PM Surya Ghar: India's Solar Revolution—Muft Bijli Yojana crosses milestone of 10 lakh installations. Ministry of New and Renewable Energy, Government of India. Details available at <https://www.pib.gov.in/PressReleaseIframePage.aspx?PRID=2111106>
12. Shah, K. 2021 (June). A renewed push on solar-powered irrigation would accelerate India's energy transition. Institute for Energy Economics and Financial Analysis. Details available at <https://ieefa.org/wp-content/uploads/2021/06/Solar-Powered-Irrigation-Would-Accelerate-Indias-Energy-Transition_June-2021.pdf>

13. Chaubal, J. 2025 (17 May). India sees a future making solar panels for itself, and maybe the world. *Deccan Herald*. Details available at <https://www.deccanherald.com/india/india-sees-a-future-making-solar-panels-for-itself-and-maybe-the-world-3-3545307>
14. Ministry of New and Renewable Energy. *n.d.* Production Linked Incentive (PLI) Scheme: national programme on high efficiency solar PV modules. Government of India. Details available at <https://mnre.gov.in/en/production-linked-incentive-pli/>
15. Ministry of New and Renewable Energy. *n.d.* Wind overview. Government of India. Details available at <https://mnre.gov.in/en/wind-overview/>
16. Ministry of New and Renewable Energy. 2025 (21 July). Programme/scheme-wise cumulative physical progress as on 30 June 2025. Government of India. Details available at <https://mnre.gov.in/en/physical-progress/>
17. Hazarika, G. 2023 (8 December). MNRE unveils revised policy for repowering wind power projects. Mercom India. Details available at <https://mercomindia.com/mnre-unveils-revised-wind-power-projects>
18. Ministry of New and Renewable Energy. (2025, September 4). Offshore Wind. Government of India. https://mnre.gov.in/en/off-shore-wind/
19. U.S. Department of State. *n.d.* U.S.–India: Civil Nuclear Cooperation. Details available at <https://2001-2009.state.gov/p/sca/c17361.html>
20. Boese, W. 2008 (October). NSG, Congress approves nuclear trade with India. Arms Control Today. Details available at <https://www.armscontrol.org/act/2008-10/nsg-congress-approve-nuclear-trade-with-india >
21. PTI. 2024 (20 September). India's third home-built 700 MWe nuclear reactor achieves criticality. *The Economic Times*. Details available at <https://economictimes.indiatimes.com/industry/energy/power/indias-third-home-built-700-mwe-nuclear-reactor-achieves-criticality/articleshow/113522142.cms>
22. Press Information Bureau. 2025 (28 April). Road map for development of nuclear power generation. Ministry of Power, Government of India. Details available at <https://www.pib.gov.in/PressReleasePage.aspx?PRID=2124961>

23. Press Information Bureau. 2024 (4 March). PM witnesses the historic 'Commencement of Core Loading' at India's first indigenous Fast Breeder Reactor (500 MWe) at Kalpakkam, Tamil Nadu. Ministry of Power. Government of India. Details available at <https://www.pib.gov.in/PressReleaseIframePage.aspx?PRID=2011347>
24. Press Information Bureau. 2025 (3 February). Nuclear Power in Union Budget 202526. Ministry of Atomic Energy, Government of India. Details available at <https://www.pib.gov.in/PressReleasePage.aspx?PRID=2099244>
25. Press Information Bureau. 2023 (25 October). 1.5 Billion Tonne Coal Production Likely by 2029-30. Government of India. Details available at <https://www.pib.gov.in/PressReleasePage.aspx?PRID=1984999>
26. CIL Navikarniya Urja Limited. 2022. *Annual Report and Accounts 2021–22*. Coal India Limited. Details available at <https://d3u7ubx0okog7j.cloudfront.net/documents/annual_report.pdf>
27. Ministry of Coal, Government of India. 2024 (February). Economic significance of the coal sector extends beyond energy. Press Information Bureau. Details available at <https://www.coal.gov.in/sites/default/files/2024-02/PIB2009196.pdf>
28. Times News Network. 2019 (19 June). India has just 2.4% of world's land but 18% of its population. *The Times of India*. Details available at <https://timesofindia.indiatimes.com/city/nagpur/india-has-just-2-4-of-worlds-land-but-18-of-its-population/articleshow/69848388.cms>
29. Ministry of Coal, Government of India. 2024 (February). *Annual Report 2023–24:* Chapter 3. Details available at <https://coal.gov.in/sites/default/files/2024-02/chap3AnnualReport2024en.pdf>
30. Press Information Bureau. 2023 (10 November). Coal India identifies 20 abandoned mines for pump storage projects. Ministry of Coal, Government of India. Details available at <https://www.pib.gov.in/PressReleasePage.aspx?PRID=1976167>
31. The Energy and Resources Institute. 2025 (June). *Reassessment of Solar Potential in India: a macrolevel study*. Details available at

<https://teriin.org/sites/default/files/files/Solar_Potential.pdf>

32. Ministry of New and Renewable Energy. 2023 (30 June). Development of solar parks and ultra mega solar power projects. Government of India. Details available at <https://mnre.gov.in/development-of-solar-parks-and-ultra-mega-solar-power-projects/ >
33. International Solar Alliance. *n.d.* Home. Details available at <https://isa.int/>
34. Ministry of New and Renewable Energy. 2025 (21 July). Pradhan Mantri Kisan Urja Suraksha Evam Utthaan Mahabhiyan (PMKUSUM). Government of India. Details available at <https://mnre.gov.in/en/pradhan-mantri- kisan-urja-suraksha-evam-utthaan-mahabhiyaan-pm-kusum/>
35. Ministry of New and Renewable Energy. 2025 (2 June). PM Surya Ghar: Muft Bijli Yojana [National Rooftop Solar Portal]. Government of India. Details available at <https://pmsuryaghar.gov.in/>

Chapter 5

1. Vickers, J., Peterson, D., and Randolph, K. 2020 (22 September). Cost of electrolytic hydrogen production with existing technology (DOE Hydrogen and Fuel Cells Program Record No. 20004). U.S. Department of Energy, Office of Energy Efficiency and Renewable Energy. Details available at <https://www.hydrogen.energy.gov/docs/hydrogenprogramlibraries/pdfs/20004-cost-electrolytic-hydrogen-production.pdf>
2. Press Information Bureau. 2022 (31 March). National Hydrogen Mission. Ministry of New and Renewable Energy, Government of India. Details available at <https://www.pib.gov.in/PressReleasePage.aspx?PRID=1812023>
3. Press Information Bureau. 2023 (1 August). Nearly 50 MMT per annum of CO_2 emissions can be averted through production and use of green hydrogen as targeted under National Green Hydrogen Mission: Union Minister for New and Renewable

Energy [press release]. Ministry of New and Renewable Energy, Government of India. Details available at <https://www.pib.gov.in/PressReleaseIframePage.aspx?PRID=1944758>

4. Henry, J. 2024 (15 February). How green hydrogen can accelerate the journey to a decarbonized future [blog post]. Hitachi Energy. Details available at <https://www.hitachienergy.com/us/en/news-and-events/blogs/2024/02/how-green-hydrogen-can-accelerate-the-journey-to-a-decarbonized-future>
5. The Energy and Resources Institute (TERI). 2022 (August). Achieving green steel: road map to a net-zero steel sector in India. TERI & Energy Transitions Commission—India. Details available at <https://www.teriin.org/sites/default/files/files/Achieving_Green_Steel_Roadmap.pdf>
6. Ministry of Steel, Government of India. 2024 (10 September). *Greening the Steel Sector in India: road map and action plan—Green Steel Initiative (GSI) Report*. Government of India. Details available at <https://steel.gov.in/sites/default/files/2025-03/GSI%20Report.pdf>
7. European Commission, DirectorateGeneral for Taxation and Customs Union. *n.d.* Carbon Border Adjustment Mechanism (CBAM). Details available at <https://taxation-customs.ec.europa.eu/carbon-border-adjustment-mechanism_en/>

Chapter 6

1. Ministry of External Affairs. (2021, November 2). National statement by Prime Minister Shri Narendra Modi at the COP26 Summit in Glasgow. Government of India. Retrieved July 21, 2025, from https://www.mea.gov.in/Speeches-Statements.htm?dtl/34466/National+Statement+by+Prime+Minister+Shri+Narendra+Modi+at+COP26+Summit+in+Glasgow
2. Ministry of External Affairs. (2023). G20 New Delhi Leaders' Declaration. Government of India. https://www.mea.gov.in/Images/CPV/G20-New-Delhi-Leaders-Declaration.pdf
3. Ministry of Environment, Forest and Climate Change, Government of India. (2024, December 30). India's Fourth Biennial Update Report (BUR 4) to the United Nations Framework Convention on

Climate Change. UNFCCC. https://unfccc.int/sites/default/files/resource/India%20BUR-4.pdf

4. CNET. (n.d.). States banning new gas powered cars. CNET. Retrieved July 22, 2025, from https://www.cnet.com/home/electric-vehicles/states-banning-new-gas-powered-cars/
5. Chatterjee, S. (2025, July 3). Govt floats tender for 10.9 K electric buses under PM e Drive scheme. Hindustan Times. Retrieved from https://www.hindustantimes.com/india-news/govt-floats-tender-for-10-9k-electric-buses-under-pm-e-drive-scheme-101751538474003.html
6. NDTV. (2025, August 4). Delhi Chief Minister Rekha Gupta flags off 105 new DEVI electric buses, inaugurates Narela depot. NDTV. https://www.ndtv.com/delhi-news/delhi-chief-minister-rekha-gupta-flags-off-105-new-devi-electric-buses-inaugurates-narela-depot-8774373
7. Akinbami, S., Ayeni, A. O., & Agboola, O. (2023). Biogas production from cow dung: A review of anaerobic digestion process, influencing parameters and recent innovations. Cleaner Energy Systems, 4, 100090. https://doi.org/10.1016/j.cles.2023.100090
8. Council on Energy, Environment and Water. (2025, April 21). State & Category Volume Monitor [Data dashboard]. CEEW-Green Finance Centre. https://www.ceew.in/gfc/tools_and_dashboards/electric-mobility/state-and-category-monitor
9. Clean Mobility Shift. (2025). EV Dashboard. https://cleanmobilityshift.com/ev-dashboard/
10. Bhatia, S. (2025, April 9). India's EV uptake hits 7.8% in FY25; are we on track to achieve 2030 target? ET Auto, Electric Times (Economic Times). https://auto.economictimes.indiatimes.com/news/industry/indias-ev-uptake-hits-7-8-in-fy25-are-we-on-track-to-achieve-2030-target/120092840
11. Ministry of Railways. (2023, March 15). Indian Railways to become Net Zero Carbon Emitter by 2030 [Press release]. Press Information Bureau, Government of India. https://www.pib.gov.in/PressReleaseIframePage.aspx?PRID=1907230
12. Mohan, R. R., & Kumar, P. (2019, December). Decarbonising the cooking sector [Briefing paper]. GHG Platform India; Centre for

Study of Science, Technology & Policy (CSTEP). https://www.ghgplatform-india.org/wp-content/uploads/publications/phase-3/GHGPI-PhaseIII-Briefing%20Paper-Energy%20Sector-Dec%2019.pdf

13. Ministry of New and Renewable Energy. (2024). Guidelines for PM Surya Ghar: Muft Bijli Yojana. Government of India. https://cdnbbsr.s3waas.gov.in/s3716e1b8c6cd17b771da77391355749f3/uploads/2024/04/2024041621270343 09.pdf
14. Press Information Bureau. (2022, June 24). Unnat Jyoti by Affordable LEDs for All (UJALA). Ministry of Power, Government of India. https://static.pib.gov.in/WriteReadData/specificdocs/documents/2022/apr/doc20224636501.pdf
15. PTI. (2025, January 1). India cut GDP emission intensity by 36 per cent during 2005–2020: Govt report. The Economic Times. Retrieved July 22, 2025, from https://economictimes.indiatimes.com/news/india/india-cut-gdp-emission-intensity-by-36-pc-during-2005-2020-govt-report/articleshow/116860223.cms?from=mdr
16. The Energy and Resources Institute. (2021, August). Green steel through hydrogen direct reduction: A policy brief. https://www.teriin.org/sites/default/files/2021-08/policybrief-green-steel.pdf
17. Global Cement and Concrete Association. (2025, March). Executive summary: Cement and concrete industry road map for net zero concrete. https://gccassociation.org/wp-content/uploads/2025/03/Excutive-summary-Roadmap.pdf
18. Ministry of Heavy Industries. (2023, December 19). Unstarred question no. 2717 answered in Lok Sabha: FAME India scheme [Parliamentary document]. Lok Sabha. https://sansad.in/getFile/loksabhaquestions/annex/178/AU2717.pdf
19. Ministry of Heavy Industries. (2024, December 10). Phase-III of FAME scheme: Unstarred Question No. 2448 answered in Lok Sabha (AU2448). Lok Sabha Secretariat. https://sansad.in/getFile/loksabhaquestions/annex/183/AU2448_7DstiP.pdf

Index